THE SCEPTICAL OPTIMIST

THE SCEPTICAL OPTIMIST

WHY TECHNOLOGY ISN'T THE ANSWER TO EVERYTHING

NICHOLAS AGAR

OXFORD
UNIVERSITY PRESS

OXFORD
UNIVERSITY PRESS

Great Clarendon Street, Oxford, OX2 6DP,
United Kingdom

Oxford University Press is a department of the University of Oxford.
It furthers the University's objective of excellence in research, scholarship,
and education by publishing worldwide. Oxford is a registered trade mark of
Oxford University Press in the UK and in certain other countries

© Nicholas Agar 2015

The moral rights of the author have been asserted

First Edition published in 2015
Impression: 1

Published in the United States of America by Oxford University Press
198 Madison Avenue, New York, NY 10016, United States of America

British Library Cataloguing in Publication Data
Data available

Library of Congress Control Number: 2014957565

ISBN 978-0-19-871705-8

Printed in Great Britain by
Clays Ltd, St Ives plc

For Alexei and Rafael. May they get the best out of future well-being technologies.

ACKNOWLEDGEMENTS

In the writing of this book I benefited from the advice and support of many people. The following friends and colleagues read and commented on parts or all of the manuscript: Jan Agar, Petra Arnold, Sondra Bacharach, Stuart Brock, Caroline Clarinval, Anthony Elder, Simon Keller, Felice Marshall, César Palacios, Mark Walker, and Dan Weijers. Latha Menon, ably assisted by Emma Ma and Jenny Nugee, did a great job of shepherding the book through to final publication. Dan Harding, the book's copy-editor, made a great many improvements to the text. I am grateful to anonymous referees for OUP who offered just the right combination of encouragement about the project and scepticism about my claims, and to Nicola Sangster for proofreading. Finally, I owe a special debt of gratitude to my wife Laurianne Reinsborough.

CONTENTS

LIST OF FIGURES

INTRODUCTION

Now is a time of both great optimism and great pessimism about technological progress. Our technologies are becoming more powerful. The apparent acceleration of this improvement has led to forecasts of colonies on Mars and cures for cancer. But the increase in the power of our technologies seems also to be increasing the magnitude of their messes. The technological errors of past generations might bring crop failures or military defeats. The errors of our current global civilization threaten human extinction. During the Cold War we lived in fear of nuclear apocalypse. The turn of the millennium saw this fear joined in the popular imagination by climate change, a potentially catastrophic legacy of the Industrial Revolution.

This book seeks to take the long view of technological progress. My focuses are not the consequences—good or bad—of particular technologies, but rather overall trends in the effects of technological innovation on human societies. The concept of subjective well-being, an increasingly important focus of work in psychology and policy studies, serves as my navigational guide. Put simply, the book's central question is: How should we expect technological progress to affect human well-being? We must look past the welter

1

of new social networking technologies, monoclonal antibody medical therapies, and self-driving automobiles to detect long-term trends in the effects of technological progress on well-being.

How do I propose to investigate the long-term effects on well-being of technological change? A long view of technological progress would seem to require comparisons of the well-being of people living at different times possessing different technologies. We lack representative samples of first-century Romans and twenty-third-century New Yorkers to quiz about their attitudes towards their ages' technologies. To take the place of these missing survey data I offer a plausible evolutionary conjecture about how our well-being responds to different levels of technology. This conjecture places limits on the long-term improvements of well-being that can come from technological progress.

An awareness of trends in the effects of technological change on well-being reveals that we are subject to a technology bias—a tendency to overstate the benefits brought by technological progress. The technology bias makes us reckless in the pursuit of these benefits. I will show how overcoming our technology bias permits a clear-eyed pursuit of more powerful technologies that is alert to relevant dangers.

Some thinkers I call radical optimists seek to encourage and amplify the technology bias. They find inspiration in an apparent acceleration of the pace of technological improvement. The radical optimists predict commensurate improvements of subjective well-being. For example, Matt Ridley expects an imminent rise in living standards 'to unimagined heights, helping even the poorest people of the world to afford to meet their desires as well as their needs'.[1] I argue that radically optimistic expectations of technological progress are mistaken.

Technological progress enhances human well-being in a paradoxical way. This paradox of technological progress concerns the translation of individual benefits from new technologies into benefits for the societies to which the individuals belong. New cancer therapies and smartphones bring significant rewards to individuals. But these benefits do not carry over, in their entirety, to the societies of which these individuals are members. Suppose that increasingly powerful technologies significantly boost the well-being of every individual exposed to them. These do not produce commensurate increases in average well-being.

There is something deeply odd about the idea that technological progress can boost the well-being of every individual in a society without much affecting the average well-being of that society. I argue that this puzzling result is produced by two phenomena.

I call the first of these phenomena *hedonic normalization*. This is the propensity for human beings to form goals that are appropriate to the environments they experience as they come to maturity. I offer an evolutionary rationale for hedonic normalization. The psychological and emotional mechanisms of subjective well-being are prominent features of human minds. They are likely to have evolved to respond to variation in the environments inhabited by humans. Humans can be hedonically normalized to arctic or to desert environments. So too, we can be hedonically normalized to different levels of technology. There is much about first-century Romans and twenty-third-century New Yorkers that we cannot know. But it's reasonable to suppose that they are hedonically normalized to their different environments, and that differences in technology will account for much of the difference between first-century Rome and twenty-third-century New York.

The second important phenomenon is *replacement*. Over time, members of a society in which there is technological progress die. Birth adds new members to that society. We see a replacement of individuals who have accrued significant hedonic benefits from technological progress by individuals who are hedonically nor-malized to the technologies that they experience as they mature. The combination of hedonic normalization and replacement means that the gradient of a line representing increases in the average well-being of a society resulting from technological pro-gress is significantly less steep than the gradient of a line repre-senting improvements in individual well-being. If I were suddenly to find myself transported to the bridge of the Starship Enterprise, then I would be extremely impressed by the technologies on display. I would promptly report to Dr McCoy for a cure for my diabetes, something that would significantly boost my well-being. But this is not how people born two centuries hence will experi-ence these technologies. They will be hedonically normalized to the technologies of the twenty-third century. We can hope that diabetes will be the kind of distant memory for most of them that bubonic plague is for most of us.

The effects of the paradox of progress are apparent in compari-sons of the well-being of people living at different times and benefitting from the technologies of their particular historical eras. Modern Romans have technologies vastly more powerful than those of ancient Romans. Ancient Romans are unavailable to participate in surveys of subjective well-being. But it's reason-able to suppose that modern Romans are not happier than ancient Romans to the extent of the difference in the power of their technologies. Modern Romans who have the best access to their era's technologies are probably only somewhat happier than ancient Romans who had the best access to their era's technologies. I argue that this much can be inferred from the evolutionary design of the psychological and emotional mechanisms of subjective well-being.

A description of the paradox of progress permits us to draw some conclusions about how best to manage technology's dangers. Technological progress is less valuable than is commonly believed in societies that currently occupy the leading edge of technology's advance. Appreciating this should help us to mitigate many of the dangers of progress.

An outline of the book

Chapter 1 introduces radical optimism: the view that technologies are becoming more powerful at an ever-increasing pace and that this increase has corresponding implications for human well-being. I describe the concept of subjective well-being which I use to challenge radical optimism. We will see that radical optimists exaggerate the benefits of technological progress.

An idea central to radically optimistic claims about the worth of technological progress is that of exponential improvement. Exponential graphs, with their characteristic initial slow growth depicting very modest improvement and subsequent rapid growth, are the gestalt of technological progress in the early twenty-first century. Chapter 2 reveals the backstory behind exponential technological improvement. The most cited examples of exponential technological improvement come from information technology. Is this pattern specific to information technology, or should we expect all technologies to be improving exponentially? I argue that exponential improvement is restricted to technologies that exhibit the property of reflexive improvement. Advances in a reflexively improving technology are themselves instrumental in facilitating later improvements. This need not mean that technologies whose improvement is currently not reflexive must be denied the advantages of exponential progress. Rapid exponential progress is infectious. It tends to spread from information technologies to other technologies. When it does so, it tends to accelerate progress in those technologies. We see this infectiousness in the increasingly important contributions of information technology to progress in medicine; surgeries are now computer assisted, computers analyse vast sets of data about people who do or don't get certain diseases for potential causes invisible to unenhanced human intuition. We should expect these contributions of information technology to further accelerate progress in medicine. This vindication of a law of exponential technological progress leaves plenty of scope for meaningful decisions about its pace and direction. The law is conditional. It assumes that users and inventors of technology will make certain choices in respect of it. The law does nothing to prevent us from relating differently with our technologies. If we less straightforwardly satisfy the conditions for exponential progress, it follows that our technologies will progress more slowly.

What are the consequences of accelerating progress for subjective well-being? Chapter 3 addresses two obstacles that seem to prevent more powerful technologies from enhancing well-being. Some commentators observe that the technologies we have good reason to think will make us happier, in fact make us less happy. Better medical technologies have created the phenomenon of the 'worried well'. New social networking technologies seem to produce status anxiety. The second obstacle arises in respect of the phenomenon of hedonic adaptation. The acquisition of a new smartphone produces an immediate boost in subjective well-being. But research on hedonic adaptation seems to suggest that we soon find ourselves either no happier or not much happier than before. I argue that neither of these obstacles prevents more powerful technologies from greatly improving the well-being of the individuals who come into contact with them. If technological progress is accelerating, it seems to follow that we should see very significant improvements in well-being.

Chapter 4 describes a new paradox of progress. The improvements of individual well-being that result from technological progress fail to translate into equivalent improvements for the societies to which those individuals belong. Improvements of subjective well-being resulting from technological progress are crucially different from other technological benefits. Most benefits of technological advances accumulate intergenerationally. If you invent a new technology for threshing corn, you pass on to your children both the technology and its effects on food stores. They receive these benefits in an undiminished form.

This book introduces the idea of hedonic normalization to explain an apparent barrier to the intergenerational transmission of the hedonic benefits of new well-being technologies. It should be viewed as an intergenerational equivalent of hedonic adaptation. Hedonic adaptation explains how individuals respond to the benefits they receive from new technologies. Hedonic normalization applies to the transmission of hedonic benefits from one generation to the next. A mother does not pass on to her daughter the hedonic benefits she has accumulated in her lifetime. Her daughter's well-being is normalized to the technologies she experiences as she matures. It follows that technological progress tends to boost subjective well-being, but to an extent far inferior to that forecast by the radical optimists.

Chapter 5 proposes a way to procure the benefits of technological progress while minimizing its risks. What is needed is a change in our attitude towards technological progress. We must begin to replace our current largely competitive approach with an approach that emphasizes cooperation. This may sound like wishful thinking. I propose one way to realize a collaborative approach to technological progress. We should conduct experiments in technological progress. These experiments would make use of the existing variation in the views of the citizens of technologically advanced societies to test different varieties of progress. I show how this can occur in respect of the potentially very beneficial but also dangerous technologies of nuclear power and agricultural biotechnology.

High on the list of priorities professed by radical optimists is an end to poverty. Chapter 6 describes the technological solution to poverty. According to the radical optimists, the harms of poverty result from material scarcities. For example, poor people do not have sufficient food, safe drinking water, or access to healthcare and education. Technological progress should, according to them, end poverty by replacing these scarcities with abundances. I argue that this understanding of poverty is shallow. Properly understood, poverty is not a consequence of material scarcities. It therefore cannot be addressed by using technology to substitute material abundances for material scarcities.

Chapter 7 addresses the issue of how to make choices about the pace of technological progress. I consider two pointers for decisions about the pace of progress. First, we should acknowledge that technological progress exhibits diminishing marginal value.

Speeding up the pace of technological progress tends to reduce the hedonic value of the goods it provides. A consequence of the diminishing marginal value of technological progress is that the faster we progress technologically, the more likely we are to correctly grant competing priorities precedence over the promotion of progress. Is there a minimum tempo of progress that we should try especially hard to achieve? I argue that we should attempt to maintain subjectively positive technological progress. A society achieves subjectively positive technological progress when technological improvement gives its members a justified confidence that it will address certain of society's outstanding problems.

Declarations that technological progress is simply good or bad may be effective as rallying calls, but they do not offer advice informed by a proper appreciation of how new technologies affect well-being. In the pages that follow, we shall see how claims about the benefits and dangers of progress can be appropriately balanced against each other, rather than being shouted as contending slogans. I describe how societies can jointly pursue an approach to technological progress—motivated by its benefits, but alert to its dangers.

1

RADICAL OPTIMISM AND THE TECHNOLOGY BIAS

Technological progress has given us many wonderful things. It has provided antibiotics and insulin injections. It has produced jet-liners that carry us around the world. It has created the Internet and given us the means to connect ourselves to it from almost anywhere on the planet.

It's not all good. Technological progress has also brought global warming and thermonuclear bombs. Uncertainties about these and other problems give an impression of precariousness. The world that technological progress has made can seem like a house of cards. In times gone by, a poorly conceived or implemented technology might lead to a famine or render a people unable to resist an invading army. The stakes are now much higher. The varieties of technological progress that enabled the Industrial Revolution are causing a climate crisis that may make large parts of the planet uninhabitable. The global store of weapons of mass destruction places the human species just a few very bad decisions away from extinction.

This book addresses a looming change in the relationship between human beings and the tools furnished by technological progress. Technologies are becoming more powerful at an accelerating pace. This quickening pace has generated an extreme optimism about technological progress. According to its proponents, accelerating technological progress will produce entirely unprecedented benefits. It will end poverty, disease, and ignorance. It will bring near limitless supplies of energy and quantities of manufactured objects. Things are getting better—*much* better. The advocates of technological progress are not blind to its problems. But they expect the quickening pace of progress to deliver prompt solutions to these problems of progress. Tomorrow's technological progress will clean up the mess made by today's technological progress.

I call this idea about technological progress *radical optimism*. Radical optimism comprises two claims. First, there's a claim about the pace of technological progress. The pace of improvement of our technologies is quickening. Each purchase of a mobile phone or a laptop computer offers a reminder of this acceleration. Second, there's a claim about the value of that progress. Radical optimists think that progress brings benefits that correspond to the pace and magnitude of technological improvement. They expect very great benefits from accelerating technological progress.

I argue that radical optimism encourages and exaggerates a *technology bias*, a systematic and often unwarranted preference for technological solutions to problems. The bias leads us to overstate the benefits of technological progress. It prevents us from recognizing that many of our problems are not eligible for resolution by the invention of new technologies. The technology

11

bias reduces our awareness of the dangers of progress. Today the bias exercises the strongest influence over the thinking of the citizens of societies at the leading edge of technological progress. We see it in the plethora of recent works celebrating the benefits of exponential technological progress. We see manifestations of the technology bias in our tendency to view Apple and Google executives not merely as purveyors of neat gadgets but seekers of solutions to our species' most challenging problems. The technology bias confers the status of sage on the IT guy.

Does technological progress increase subjective well-being?

How should we evaluate radical optimism? This book considers technological progress in terms of its contributions to human well-being. Does technological progress tend to improve well-being? Will accelerating technological improvement produce societies with ever-higher levels of well-being? Or will the quickening pace of technological progress produce environmental disasters that will lead global well-being to plummet?

Our focus on these questions about technology and well-being is broad—we are interested in a wide range of technologies that improve well-being in a variety of ways. There are pressing questions about the beneficial or harmful effects of specific technologies. Many of the questions that should be asked about nuclear power plants differ from those that should be asked about social networking technologies. This book looks beyond individual technologies to a heterogeneous category of technologies that share the purpose of sustaining or improving human well-being. I will use the label *well-being technologies* to describe the diverse collection of technologies whose direct purpose is either to improve well-being or to protect against influences that would reduce it. The category of well-being technologies includes technologies that prevent, treat, or cure diseases, technologies that enable social exchanges between people, technologies that transport people over long distances, technologies that increase the yields of crops or facilitate healthful nutrition, technologies that protect humans from the vagaries of nature, and so on. According to radical optimists, these varied well-being technologies are subject to the same law of technological progress. It's this accelerating improvement that supports predictions of significantly enhanced future well-being.

A broad focus on technological progress can convey truths missed by investigations of the specifics of given well-being technologies. Being overly focused on certain well-being technologies can distract attention from a general understanding of how technological progress improves well-being, much in the way that being overly focused on the trees distracts attention from truths about the forest. We need to take a step back from the details of the technologies themselves to become properly aware of the implications of accelerating technological progress.

It's clear that human interests in technology extend beyond the promotion of well-being. The advancement of scientific knowledge is an important value. The Large Hadron Collider, located on the Franco-Swiss border, is technology built to help answer some of the deepest scientific questions about the origins and nature of the universe. Its discoveries have greatly enhanced the well-being of some physicists. But it would be a mistake to describe the Collider's purpose as the enhancement of human well-being. The protection and promotion of well-being is, nevertheless, an important purpose of human technologies and technological progress. Contributions to well-being have some claim to be the most important ways in which any form of progress, technological or other, should be evaluated. It is difficult to imagine an alternative end so important that its pursuit justifies systematically reducing well-being. Moreover, it is difficult to imagine an alternative end so important that its pursuit would justify forgoing an opportunity to significantly boost well-being.

One reason why well-being is an especially apposite focus for a sceptical discussion of radical optimism is that its advocates make their case for accelerated technological progress in terms of what they see as its propensity to significantly boost well-being. I will argue that predictions of increased well-being from technological progress are vastly overstated. Once we understand how new technologies boost well-being we will reject forecasts of increases of well-being that correspond to the power of future technologies.

Choosing to locate the concept of well-being at the centre of my discussion invites difficulties. The concept of well-being has consumed the efforts of some of our greatest philosophers. In spite of this, or perhaps because of it, there is no consensus on how to define the concept, and therefore what it means for an individual's well-being to improve. This is a problem for those interested in how technological progress might boost or decrease well-being. Without a good understanding of well-being, our pronouncements about the value of technological progress lack a secure foundation.

To circumvent some of these disputes about words and word meanings, I conduct my discussion in terms of the concept of *subjective well-being*.[2] The subjective aspect of this concept makes it a fair approximation of our concept of happiness. We would call people with high subjective well-being happy and people with low subjective well-being unhappy. Subjective well-being has recently assumed great significance in policy studies. Social scientists are currently using the concept to get a better understanding of the human effects of a variety of social changes. Some of the concept's advocates propose subjective well-being as a replacement for, or a supplement to, economic measures of governments' achievements.

A person's subjective well-being refers to how she experiences living her life. The concept comprises two measures. There is a cognitive measure that assesses life satisfaction. It draws on judgments that we make about how well our lives are going. How good do we judge our lives to be? Questionnaires are a principal tool in the investigation of the cognitive component. The second measure assesses our affective states. The aim here is to gauge the flow and balance of positive and negative feelings or emotions. How good do our lives feel? Some investigations of subjective well-being's affective component involve devices with alarms programmed to ring periodically. When the alarm goes off, study participants write down a number that indicates how happy, in that very moment, they feel. Those who tend to report more positive feelings and emotions are interpreted as, in this respect, having higher levels of subjective well-being than are those who report fewer such feelings and emotions. The two measures can give different answers. Individuals who report high levels of life satisfaction may nevertheless tend to report comparatively few positive experiences. Individuals who proclaim themselves to be quite dissatisfied with how their lives are going may nevertheless report many pleasant experiences and few unpleasant ones.

The great advantage of the concept of subjective well-being is that it refers to something that can be measured. It's not so surprising to learn that the loss of a job can reduce well-being. But the concept of subjective well-being permits us to quantify that loss. Its constituent measures enable us to compare the efficacy of alternative remedies for the hardship of job loss.

The concept of subjective well-being is not immune to philosophical challenge. There are some cases in which it offers affirmative verdicts on human lives that are intuitively worthless and negative verdicts on human lives that are intuitively worthwhile. These counterexamples show that there is not a perfect fit between our intuitive understanding of well-being and the concept of subjective well-being. It might seem that building an account of the human effects of technological progress on a potentially false theory of well-being is a bit like using rotten timber in the foundations of a house. A mistaken theory of well-being fails to justify conclusions about technological progress just as rotten foundations fail to support a house.

One problem is that this entire book could be dedicated to the task of defining well-being and we would still not have convinced all of the philosophical doubters. In dealing with philosophical uncertainties I propose to take the lead from science. Sometimes uncertainties in measurement force scientists to make do with estimates. In a perfect world those seeking to forecast climate change would appeal to exact measures of the quantity of carbon dioxide currently in the atmosphere. Since we cannot count every last molecule of carbon dioxide, we must rely on estimates that we hope are sufficiently close to the truth so that our conclusions will successfully apply to the real world. I think that we can treat the concept of subjective well-being as an estimate of a philosophically satisfactory theory of well-being. Even if not correct in every detail, the concept of subjective well-being should be close enough to the truth about human well-being—the concept that we hope philosophers will eventually agree on—for this book's conclusions to successfully describe how technological progress affects well-being.

The alternative of awaiting philosophical consensus on what it means for well-being to increase is effectively resigned to never making practical use of the concept. If technological progress makes a difference to the judgments we make about how our lives are going, or the feelings and emotions that we experience, then an investigation guided by the concept of subjective well-being should reveal them. It's reasonable to suppose that the concerns about technological progress highlighted by the concept of subjective well-being will be expressible in the terms of some more philosophically sophisticated analysis of well-being.[3]

Radically optimistic forecasts

The radical optimists think that the increasing steepness of the curve of technological progress will dramatically enhance human well-being. Their buzzwords are *optimism*, *abundance*, and *infinite*.

Consider some recent expressions of radical optimism. Matt Ridley's 2010 book *The Rational Optimist: How Prosperity Evolves* presents a picture of our near future in which 'the pace of innovation will redouble and economic evolution will raise the living standards of the twenty-first century to unimagined heights'.[4] These immense benefits of progress will be enjoyed by all. In his 2012 book *The Beginning of Infinity: Explanations that Transform the World*, the theoretical physicist David Deutsch advances a principle of optimism according to which there are no insuperable problems, no barriers to progress that cannot be overcome.[5] According to Deutsch, we are universal explainers, capable of propounding theories with infinite reach. The limitless reach of our explanations leaves no problem without a solution that cannot be found by humans equipped with the tool of science. The potentially infinite knowledge of science brings with it the potentially limitless know-how of technology.

Advances in information technology take centre stage in many expressions of radical optimism. In their 2013 book *The New Digital Age: Reshaping the Future of People, Nations, and Business*, the Google executives Eric Schmidt and Jared Cohen urge that 'The best thing that anyone can do to improve the quality of life around the world is to drive connectivity and technological opportunity.'[6] They say 'future connectivity promises a dazzling array of "quality of life" improvements: things that make you healthier, safer and more engaged'.[7] The Internet is the hero of Byron Reese's 2013 book *Infinite Progress: How the Internet and Technology Will End Ignorance, Disease, Poverty, Hunger, and War*. Reese predicts that the misfortunes and failings of his subtitle are 'all about to vanish, courtesy of the Internet and its associated technologies'.[8] Ramez Naam is another exponent of the concept of infinity. His 2013 book *The Infinite Resource: The Power of Ideas on a Finite Planet* shows how new technologies can address problems that result from the apparent finitude of our natural resources.

Among the books that place the concept of abundance at the heart of their advocacy of technological progress is *Abundance: The Future Is Better Than You Think*, a book written by Peter Diamandis and Steven Kotler. Diamandis founded the X-Prize

Foundation, an organization responsible for incentivized prize competitions whose purpose is to bring about 'radical breakthroughs for the benefit of humanity'.[9] Diamandis and Kotler want to end scarcity. They look to technological progress to realize 'a world of nine billion people with clean water, nutritious food, affordable housing, personalized education, top-tier medical care, and nonpolluting, ubiquitous energy'.[10] The title of K. Eric Drexler's 2013 book *Radical Abundance: How a Revolution in Nanotechnology Will Change Civilization* seems to indicate a desire to outdo Diamandis and Kotler.[11] Drexler describes in some detail the means by which nanotechnology can end scarcity. In the future, as perceived by Drexler, miniature factories churn out near endless supplies of any of the products that we desire. These mini-factories neither create waste nor require toxic heavy metals.

How should we prioritize technological progress?

It is important to avoid too simplistic a view of the decision about technological progress we must make. We are not choosing whether to endorse or reject technological progress. This book advocates no sweeping rejection of technology or technological progress. That would be absurd. Humans are a technological species. Hand axes and cooking fires were prominent features of the environments that shaped human evolution. Progress in technology is a universal feature of human societies too. Humans everywhere and at every time have sought to rearrange parts of their environments to improve their lives. This fact is as true of contemporary inhabitants of North Korea as it is of residents of Silicon Valley.

But we should not be tricked into thinking that this leaves only the option of uncritical acceptance. In Chapter 2 I will consider the claim that accelerating technological progress is a consequence of a law of exponential improvement. The idea that technological progress conforms to a law may seem to suggest that we have no opportunity to make meaningful decisions about it. But intervening in a law of technological improvement is not like falsifying one of Newton's laws of motion. The former is a conditional law that assumes the contributions of appropriately motivated human users and inventors of technology. As we shall see in Chapter 2, we can collectively influence the pace and direction of technological progress by deciding how and to what degree we satisfy the conditions for its acceleration. When humans choose to engage with technology in different ways, they can alter the pace and direction of its advance.

Some of the most difficult decisions made by the leaders and citizens of modern democracies involve the setting of priorities. Often, we are not asked to decide whether a given activity is good or bad. Rather, we are dealing with activities that we accept as intrinsically worthwhile. We need to know *how* good each activity is. Such decisions permit us to properly locate them among our priorities. Consider two alternative policies, one that should improve the quality of early childhood education and another that predictably reduces the incidence and severity of coronary artery disease. The effects of both policies are clearly good. With limitless time and resources we'd aim for a lot more of both. The finitude of time and recourses necessitates difficult decisions about when to forgo the opportunity to pursue one goal so as to better pursue another.

Speaking in 1953, President Dwight Eisenhower demonstrated a good understanding of trade-offs when he said of the US military budget: 'Every gun that is made, every warship launched, every rocket fired signifies, in the final sense, a theft from those who hunger and are not fed, those who are cold and are not clothed.'[12] Eisenhower was a former soldier who was president during the Cold War. We can assume that he had a good understanding of the importance of maintaining military readiness. But he understood some of the costs of that readiness. Money spent on guns and rockets is unavailable to be spent on feeding hungry people, clothing cold people, or for that matter on educating them, or treating their illnesses.

Consider this reasoning applied to technological progress. Just as a decision to build more guns and fire more rockets tends to reduce the resources we can commit to feeding people, so too, the resources we commit to technological progress cannot be put to other uses. Here's a danger of radical optimism. We know that technological progress is good and important. We know that it supports and enhances human well-being. But we need to know *how* good and *how* important technological progress is. We should be wary of influences that lead us to overstate its value. This book argues that the radical optimists overstate the value of technological progress. This overstatement gives technological progress an undeserved pre-eminence over alternative priorities. Resources and effort committed to accelerating technological progress are not available for important endeavours properly viewed as independent of technological progress.

An additional feature of radical optimism further exaggerates the value of technological progress. Radical optimists tend to subscribe to an *instrumentalism* about technological progress.

I will understand instrumentalism as the claim that all, or very many, of our problems are best addressed by encouraging technological progress. Writing about the Internet and other digital technologies, Evgeny Morozov challenges an ideology of 'technological solutionism'. This solutionism recasts 'all complex social situations either as neatly defined problems with definite, computable solutions or as transparent and self-evident processes that can be easily optimized—if only the right algorithms are in place!'[13] For technological solutionists there's almost always an app for that. Problems like obesity or childhood poverty await their digital fixes.

Instrumentalism is a broader thesis than technological solutionism. It concerns technological progress in general, and forms a particularly potent combination with accelerating technological progress.[14] Instrumentalism leads us to expect imminent solutions to problems that are currently out of the reach of technology. The Montgolfier brothers faced the problem of how to transport humans into the sky. Their technological solution was the hot air balloon. According to its advocates, accelerating progress will transform problems long out of the reach of technology into technological problems with technological solutions. Radical optimists expect technological solutions to the problem of global poverty, the problem of how best to educate our children, and the problem of how to fix the damage done to our planet by past technological progress. Homer Simpson memorably describes alcohol as 'the cause of, and solution to, all of life's problems'. Perhaps something like this is true of technological progress. It may be the cause of many of the world's problems but so long as it solves them too then that's not so bad. Radical optimists present technological innovation as the master key to a better world.

Instrumentalism is mistaken. The pursuit of technological progress must not serve as a proxy for improving social justice or ending poverty. The right new technologies can be important contributors towards these ends. But they should be recognized as playing a strictly subsidiary role. As we shall see, the idea that technological innovation should play the dominant role in our collective response to the problem of poverty, injustice, and environmental degradation tends to exacerbate problems rather than solve them. The technology bias inclines us to place too much value on technological progress. Radically optimistic appeals to accelerating technological advancement threaten a further overvaluation. We should expect a further skewing of the priorities of technologically advanced societies.

The preceding paragraphs suggest that fast-improving technologies may not live up to some of our expectations. They are no master key to a better world. But there are also distinctive dangers in an over-reliance on technology. The phenomenon of anthropogenic climate change has sharpened our awareness of the dangers of technological progress. We are only now coming to terms with some of the costs of the Industrial Revolution. An overvaluing of technological progress leads us to pursue it in a way that we should acknowledge as reckless.

Consider the following analogy inspired by Homer Simpson's advocacy of alcohol. Both heavy and moderate drinkers enjoy alcohol. The overvaluation of alcohol by some heavy drinkers leads them to pursue it in a way that they should recognize as dangerous. The lesser valuation of the moderate drinker permits alcohol to be enjoyed in a way that is safer and does not threaten other important commitments. Radical optimists are like the heavy drinkers. They are intoxicated by the potential benefits of technological progress. I will argue that we can better see how to reduce the many risks of technological progress once we appreciate that its benefits are systematically oversold. The mere act of scaling back the value of technological progress indicates how to avoid its many hazards and pitfalls. I will describe a more cautious approach to technological progress enabled by freeing ourselves of our technology bias.

Concluding comments

We now have some idea of the radically optimistic claim that accelerating technological progress can significantly enhance well-being. The next chapter explores a law that is often invoked to explain the increasing momentum of technological progress. According to the law of exponential improvement, the dazzling advances that seem to be defining features of our age are not by-products of a passing fascination with technology. They are direct consequences of a law of progress that causes our technologies to become more powerful at an exponential pace. The law of exponential improvement supports the confidence of radical optimists about the increasing benefits of technological progress.

2

IS THERE A LAW OF TECHNOLOGICAL PROGRESS?

Here are two snapshots of progress in well-being technologies. One is a great advertisement for radical optimism. The other seems to be a stinging rebuff.

First, there's the mobile phone. Few people who have any experience of mobile phones can fail to be impressed by the speed and magnitude of progress in this technology. We have gone from a time in the early 1980s when mobile phones weighed a ton, cost a bomb, and frequently dropped calls, to the cheap, sleek smartphones of the second decade of the twenty-first century, complete with connections to the Internet, access to GPS satellites, and more computing power than a several million dollar Cray Supercomputer from the 1970s. This progress has such a prominent place in the collective consciousness that the mere appearance of the brick-like mobile phone used by Gordon Gecko in the iconic 1987 movie *Wall Street* was an effective sight gag when returned to him in the 2007 sequel *Wall Street: Money Never Sleeps*. The near future will bring devices that make the purportedly twenty-third-century communicators of *Star Trek* seem increasingly primitive.

And then there's cancer. The story of the search for a cure for cancer seems to have been one of serial disappointments. President Richard Nixon declared what must have seemed to him a very winnable war on cancer in 1971. In spite of many proclamations of an imminent final victory, the war remains, over forty years later, stubbornly unwon.[15] This is not to say that there has been no progress. It was better, in general, to receive a cancer diagnosis in 2007 than in 1987. We are getting better at preventing, detecting, and treating cancer. But this improvement doesn't seem to be accelerating. Women who learn that they have inherited a version of the BRCA1 gene that elevates their risk of breast cancer feel no entitlement to complacency on account of an accelerating improvement of cancer therapies that should inevitably prevent any problem before it manifests.

This chapter investigates evidence for a law of exponential technological improvement. A law which directs that technological progress is exponential offers crucial support for radically optimistic forecasts.[16] While there is strong evidence for such a law in information technology, it is less apparent outside of the realm of computers. Exponentially improving information technologies do make important contributions to human well-being. But if exponential improvement were to be strictly limited to information technology, it would justify no optimism about many of the technologies that make the greatest contributions to improving human well-being. Most people would gladly trade any number of advances in social networking technology for a proper cure for cancer.

I will argue that exponential technological improvement is infectious. It tends to spread from rapidly advancing information technologies to other technologies. We are currently witnessing a transfer of rapid exponential improvement from information technology to a broad range of well-being technologies. Where exponential technological improvement gains a foothold the magnitude of its influence tends to increase. This conclusion justifies the confidence of radical optimists about the pace of technological progress. It means that those who are concerned about radical optimism's implications for society should not take refuge in predicted failures of technological change to deliver accelerating improvements.

Moore's Law, Kryder's Law, and exponential technological improvement

The most celebrated example of exponential technological progress comes from information technology. Moore's Law, named for the Intel Corporation co-founder Gordon Moore, states that the power of integrated circuits, key components of modern computers, doubles approximately every two years. You experience the power of Moore's Law every time you replace your personal computer with a newer model.

The graph that describes Moore's Law features an initial slow growth part that corresponds to the modest, hard-won improvement of the early days of integrated circuits (see Fig. 1). This is followed by a rapid-growth part representing the manifest and marked improvement in the power of the processors that find their way into the computers of the early twenty-first century. In contemporary discussions of technological progress, graphs of exponential growth have acquired an iconic status of the type possessed by the intersecting lines of supply and demand graphs in introductory economics texts. The law of exponential

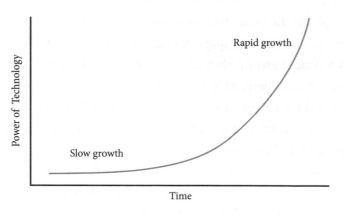

FIG. 1 The exponential improvement of technology.

technological improvement purports to explain why this trend is no ephemeral pattern peculiar to our technology-obsessed age. It seems to justify an optimism about technologies that become ever more effective at promoting and enhancing human well-being. Those who advocate exponential progress point to evidence that we are entering the rapid-growth parts of the curves of improvement of a wide variety of technologies. Technological progress in the past has seemed halting and arduous. From now on, we will experience an accelerating rush of new and improved technologies as the exponential curve of improvement becomes increasingly vertical. There are corresponding consequences for the benefits produced by our technologies. We are entering a future in which the technologies that entertain us, protect us from disease, and transport us around our world and beyond are becoming exponentially more powerful.

The central idea of exponential growth is deceptively easy to state. It occurs when a mathematical function describes a pattern of growth that is proportional to its value at any given time. As the value of the function increases, there is growth in its rate of increase. Consider one quite dramatic presentation of the power of exponential growth that takes the form of an experiment that can be done at home. At least, its early stages can. The experiment is essentially an origami-for-dummies exercise. Take an ordinary sheet of A4 paper. The sheet has a thickness of approximately 0.1 millimetres. Fold it. You should produce something with a thickness double that of the unfolded sheet—0.2 millimetres. Now fold it again. If you've been conscientious in your folding the resulting stack should be 0.4 millimetres thick. Obviously this folding exercise becomes more difficult as you proceed. In my experience it's hard to get beyond six folds. But suppose that you could fold the sheet one hundred times. How thick do you think the stack would be? Would it be as thick as a several hundred-page telephone book? Might its thickness match the height of a refrigerator? Or the height of a house? The true, but incredible answer is that its thickness would be 12 billion light years, the approximate radius of the known universe. By fifty folds the stack already measures the distance from Earth to the Sun.[17] It is an indicator of the extent to which exponential growth exceeds human expectations that, as I type these sentences, I cannot quite believe what I'm typing. How could the easily imagined act of folding a piece of paper one hundred times lead to something half as wide as the observable universe?

A common theme of radical optimists is that exponential technological improvement systematically confounds human comprehension. We tend to be caught out by an exponential curve's transition from slow growth to rapid growth. We fail to anticipate what Erik Brynjolfsson and Andrew McAfee call the 'inflexion point', which occurs when slow exponential progress in a wide range of information technologies gives way to rapid progress.[18] This collective blind spot explains mistaken predictions about computers that take the form of 'computers will never do X' where X refers to some human ability. A prominent example was the confidently expressed prediction that computers would never play chess sufficiently well to defeat the best human players. Through the 1980s and early 1990s various commentators observed the performance of chess computers and noted the huge and seemingly persistent gap between them and the best human players. Garry Kasparov, a candidate for the strongest-ever human player, was defeated in 1997 by the IBM-programmed computer Deep Blue. Kasparov seems to have been a victim of the seemingly abrupt transition of chess computers from the slow growth phase of exponential improvement to the rapid growth phase. Our exponential change blind spot ill prepares us for a world made by information technology.

Some commentators worry about an imminent end for Moore's Law as we begin to reach barriers imposed by the basic physical properties of matter.[19] Ray Kurzweil predicts that Moore's Law will come to an end by 2020. According to him, the demise of Moore's Law will enable another technological paradigm to bring exponential growth to computers, just as Moore's Law followed on from earlier paradigms. 'Each time one paradigm has run out of steam, another has picked up the pace.'[20] When Moore's Law comes to an end, the exponential growth in computer processing power will continue. Possible new paradigms involve 3D chips in which transistors are stacked rather than arranged on a single two-dimensional layer, and quantum computers which process information by the direct use of quantum mechanical phenomena.

If Moore's Law stood in isolation from other examples of technological improvement then it would be noteworthy enough. Integrated circuits are hugely important components of modern computers. They are, as a consequence, key elements of a wide range of early twenty-first-century well-being technologies. However, viewing exponential progress as limited to Moore's Law significantly undersells its significance. Exponential improvement is apparent in many other facets of information technology. For example, Kryder's Law applies to the storage of magnetic technologies. It too is unmistakably exponential. Kryder's Law specifies that the storage density and therefore the capacity of hard disks to store data doubles every thirteen months—thus achieving a rate of improvement faster than that described by Moore's Law. It has held up across changes in ways of storing data. Kurzweil documents patterns of exponential improvement in many other information technologies, for example in improvements of dynamic random-access memory (DRAM),[21] microprocessor clock speed,[22] the power of supercomputers,[23] the volume of Internet data traffic,[24] and so on.

Two questions about exponential technological progress

Moore's Law, Kryder's Law, and other laws directing exponential progress in information technology are obviously fascinating for people interested in the future of computers. But what are the implications of exponential technological progress for human well-being?

The following pages address two questions about exponential technological progress. First, we observe exponential improvement in some technologies. But is that really a consequence of a law? If exponential progress conforms to a law then we might expect it to continue. It will be no fleeting feature of our technology-obsessed age that we should expect to vanish as soon as teenagers stop finding Samsung Galaxy smartphones and Sony PlayStations cool. I will show that we can understand exponential technological progress as driven by a law.

A second question concerns how broadly a law of exponential progress applies. Is it about more than just computers? Information technologies make important contributions to a wide range of well-being technologies. But improvements in these elements do not seem to translate automatically into improvements of human well-being. Suppose that automatic back-massaging chairs were controlled by increasingly powerful computer chips. Exponential improvements in the chips need not translate into massages that are all that much more pleasurable than those provided by older designs. It would be unrealistic to expect customers to be emerging from the later chairs with feelings of satisfaction many multiples of times better than the satisfaction experienced by earlier customers.

Exponential technological improvement as a conditional law

What kind of law could a law of exponential technological improvement be? In a review of Kurzweil's 2012 book *How to Create a Mind* the philosopher Colin McGinn challenges the possibility of such a law. He views it as a mere accident of history that we should not expect to continue into the future. McGinn says:

> Kurzweil's 'law' is more likely to be fortuitous than genuinely law-like: there is no *necessity* that information technology improves exponentially

over (all?) time. It is just an accidental, though interesting, historical fact, not written into the basic workings of the cosmos. As philosophers say, the generalization lacks nomological necessity.[25]

In this context, nomological necessities are consequences of the universe's most basic physical laws. The laws of physics and chemistry are paradigms of nomological necessity. One thing is clear. The law of exponential technological improvement, if it is to be properly considered a law, must be a law of a very different type from the laws of physics or chemistry. It does not have the exceptionless universality of the laws described by Einstein's theory of relativity or Boyle's theory of gases. One could imagine a universal human disenchantment with information technology that would lead the integrated circuits of the future to be no better than those of today. This imaginative act does not seem akin to dreaming up a universe in which objects routinely contravene Einstein's theory of relativity and travel faster than the speed of light. Complete technological stasis may be improbable, but it does not seem a physical or sociological impossibility.

Such objections misunderstand what kind of law the law of exponential improvement is supposed to be. It is a conditional law. Its operation is contingent on the existence of appropriately motivated intelligent beings capable of forming plans about new technologies and implementing them. The law of exponential technological improvement is not alone in making this kind of assumption. Consider the law of supply and demand in economics. According to this law, the price of a category of goods or services is determined by an interaction between its supply and the demand for it. The law of supply and demand is a valuable tool for explaining and predicting market behaviour. It is conditional. The law assumes the existence of both agents who seek to pay as little as possible and agents who seek to charge as much as possible. There is no basic law of the universe that would prevent people from resolving to use a random number generator to set the price of any good or service. This means that the law of supply and demand has limitations. It tells us nothing about how goods are priced in a society in which each financial transaction involves the use of a random number generator. While the law fails to explain the behaviour of agents in this imagined society, its possible existence is no threat to a law of supply and demand. We must distinguish states that are consequences of a law, states that should they not obtain would falsify it, and states that the law assumes. A law simply fails to apply to states that do not satisfy its conditions, but they do not count against its soundness.

The law of exponential technological improvement assumes the attention and commitment of intelligent agents who seek improvements of the technologies that they use. This assumption is appropriate. It captures a general truth about us. Human beings at all times and in all places have sought to improve their circumstances, and one of the best ways to do this is by inventing more powerful technologies. But this widespread interest is no consequence of any law of technological progress. Moore's Law does not compel chip designers to seek more powerful integrated circuits. It merely assumes that they will.

There is nothing in the law of exponential technological improvement that prevents humans from engaging differently with technologies. Different ways of engaging may lead a society to completely fail to satisfy a condition of the law or, less extremely, they may cause the society to partially satisfy its conditions. When humans lose interest in a technology we should expect to see a deceleration in the pace of its improvement. The horse-drawn chariot is a technology that has attracted the interest of few human inventors and designers over the past millennium. It should therefore not be surprising that improvements in techniques for hitching horses to chariots have not been exponential over this period. A condition for exponential progress is not satisfied. Just as there are no intrinsic facts about gold ingots that compel buyers to pay large sums of money for them, there are no intrinsic facts about integrated circuits that compel humans to improve them at a very fast pace.

Acknowledging the law of exponential technological improvement as a conditional law gives plenty of opportunity for us to influence the speed of technological progress. We can imagine a society of hippies that seeks to satisfy none of the conditions of exponential improvement. Alternatively, we can imagine a society of *Star Trek* devotees that seeks to maximally satisfy them. The option of selectively satisfying the conditions of exponential progress and thereby achieving an intermediate pace is also available.

What went wrong with cancer?

Suppose that we grant the existence of a law of exponential technological improvement. How broadly does such a law apply?

We have little difficulty in recognizing exponential progress in information technology. But it seems less apparent in technologies that make more direct contributions to human well-being. Consider progress in medicine. Medical well-being technologies have always attracted the interest of innovators and inventors and will, presumably, continue to do so. They seem therefore to satisfy the conditions for exponential progress. How do we reconcile this claim with the disappointing nature of advances in our knowledge of how to treat disease in the later years of the twentieth century and early years of the twenty-first? Progress seems systematically to have disappointed grand expectations. The first half of the twentieth century seemed a time of very rapid progress.[26] The hormone insulin was isolated. When injected by diabetics, it turned a fatal condition into a chronic one. Antibiotics were discovered. Great killers of human beings including tuberculosis and syphilis were now cured. In the 1970s, a mass vaccination programme eradicated smallpox—1977 saw the world's last diagnosis of the disease. As recently as the year 1967, there were ten to fifteen million cases of smallpox with two million deaths.[27] There has certainly been progress in medicine since then. But we seem to lack anything of the magnitude of the isolation of insulin, the discovery of antibiotics, or the eradication of smallpox. If anything, progress seems to have slowed rather than accelerated.

The disappointments of technological progress are not limited to medicine. The entrepreneur and technology pundit Peter Thiel observes that, in our era, significant advances seem to characterize technologies that don't matter much while being absent from technologies that really do matter. In a reference to the character limit on postings to Twitter, Thiel says: 'We wanted flying cars, instead we got 140 characters.'[28] For variants of this complaint try 'We wanted a cure for cancer, instead we got [substitute the name of any recently hyped social networking technology].'

One reason that a law of exponential improvement applies so much better to progress in information technology than it does to progress in other technologies concerns the availability of evidence. The history of cancer medicine presents as a sequence of therapies ever better at extending life and alleviating symptoms.

In this respect, the history of cancer medicine resembles the history of integrated circuits. The history of integrated circuits is clearly one in which later chips are better than earlier ones. But the history of the integrated circuit provides an additional piece of information that permits us to say that progress is not only occurring, it is accelerating. We can make quantitative comparisons. A count of the number of transistors on an integrated circuit reveals a doubling approximately every two years. Hence we have accelerating progress. It's more difficult to make these quantitative judgments in respect of cancer medicines. We can say that the best treatment for chronic lymphocytic leukaemia in 2015 is better than the best treatment of 1985. But by virtue of which facts would we say that it is twice or three times as good? Suppose that the best treatment for a certain cancer in 2015 grants an average of ten extra years of life. The best treatment for that same cancer in 1985 granted, on average, an additional seven years of life. It is clear that there has been progress. But is the best treatment in 2015 three times, twice, or one and a half times better than its 1985 predecessor?

The lack of direct evidence should not lead us to reject the possibility of exponential progress in cancer medicine. We should not confuse the absence of evidence for exponential progress with evidence for its absence. Dinosaur fossils are better preserved in some rock formations than in others. It would be wrong to conclude the absence of dinosaurs from the absence of fossils in areas in which rock formations are known to be inhospitable to fossils. We can have strong indirect evidence where direct evidence is lacking. If the inhospitable rock formations occur in the midst of regions that are both good at preserving fossils and do preserve them we might have good reason to believe that dinosaurs did inhabit those areas too. We might have good reason to believe in exponential progress in cancer medicine even if we lacked a straightforward measure such as the one central to Moore's Law.

There are two pessimistic hypotheses about technologies that do not involve information technology. One is that progress occurs, but it is not exponential. In this view, progress in medicine lacks the very specific features of information technologies that lead them to progress exponentially.

Another pessimistic hypothesis allows that medical technologies are advancing exponentially. But this exponential progress fails to justify optimistic forecasts about the effects of technological progress on well-being. Suppose that the agricultural tools of medieval peasants were subject to a law of exponential technological improvement. This need not mean that things were going to get better for them anytime soon. The implements of medieval agriculture were in the slow growth phase of exponential improvement. The peasants should have expected generations of scything and threshing with technology improving at a very slow rate before the onset of the rapid growth that would swiftly bring the combine harvester. Perhaps medical technologies are still in the slow growth phase of their curves of exponential improvement.

There is some support for this idea from the history of medicine. One consistent cause of disappointment is a systematic tendency to underestimate the difficulty in successfully treating or curing disease. The dramatic advances of the early twentieth century involved diseases that turned out to be easy to treat. The lives of people with type 1 diabetes were significantly extended by the simple expedient of injecting a hormone that they lacked. There is no analogous, easily correctible deficiency in the brains of people who go on to develop Alzheimer's disease. Effective therapies for Alzheimer's are simply much harder to find than are effective therapies for diabetes. Cancer is a much more complex disease than it at first seemed. It is now clear that few cancers can be straightforwardly cut out of human bodies. Perhaps future historians of cancer will confirm that, although progress is exponential, we are currently, and will remain for quite some time, in the slow growth part of the relevant exponential curve or curves.

The idea that some technologies are improving at a pace that is slower than others because they are still in the slow growth phases of their exponential curves may not seem so bad. But it would be bad news for radical optimists if many of the technologies that matter most to human well-being are grinding through the slow growth parts of their curves. It would be a particularly cruel good news, bad news joke. The good news is that the pace of improvement of all well-being technologies is exponential. The bad news is that we have a couple of centuries to go as we trudge through the slow growth phases of the curves describing the improvement of many of the most important well-being technologies. Expect a smartphone capable of taking and sending trillion-pixel puppy photos very soon. But you still have a century to wait for treatments of multiple myeloma that are very much better that those available today.

If we had the certain data points of Moore's Law we could establish where on the exponential curves of medicine we are. Without them, we are left with our vague impressions of progress. The radical optimist Byron Reese sees the glass of medical progress as half full. He offers a list of increasingly many diseases cured or effectively treated in each decade that gives an impression of progress that is both exponential and rapid.[29] In Reese's history of medicine, centuries of slow progress in treating disease conclude with a slew of treatments and cures in the final years of the twentieth century and early years of the twenty-first. Others will give a list of the many failed attempts to cure cancer that seems to convey the opposite message. This glass half empty picture emphasizes that fundamental advances in the understanding of disease processes are required before we can hope for therapies that genuinely cure rather than merely reduce the severity of some symptoms.

The radical optimists require not only that technologies that bear directly on human well-being are undergoing exponential improvement. They want this improvement to be entering the rapid-growth part of the curve of exponential improvement. In the remainder of this chapter I consider two reasons why progress in a broad range of well-being technologies might now or soon be both exponential and rapid.

Kurzweil offers one possible reason why progress in well-being technologies should conform to the rapid exponential progress of information technology. He proposes that information technology and technologies that bear more directly on human well-being both acquire their exponential pace of improvement from a common cause: the evolutionary process. I will reject this explanation. A second possible explanation posits a kind of infectiousness of rapid exponential improvement. Well-being technologies tend to acquire rapid exponential progress from information technologies.

Kurzweil's evolutionary explanation of exponential technological progress

Kurzweil views the manifest exponential progress of Moore's Law and Kryder's Law as examples of a broad phenomenon that is not specific to information technology. He traces the exponential improvements of integrated circuits and data storage to the common cause of the evolutionary process. The law of exponential improvement 'describes the acceleration of the pace of and the exponential growth of the products of an evolutionary process'.[30] Kurzweil says 'an evolutionary process inherently accelerates... and that its products grow exponentially in complexity and capacity'.[31] Our brains and bodies are products of evolution. But so are the technologies that we use those brains and bodies to make. If the evolutionary process involves exponential improvement, then we as evolved beings transmit this exponential improvement to all of our technologies. Since all human technologies stand in essentially the same relationship with the evolutionary process we should expect them all to inherit its exponential pace. What's true of the integrated circuit should also be true of therapies for leukaemia.

That the evolutionary process can improve things is well established. Evolution turned light-sensitive patches of skin into the awesome complexity of the human eye. But what reason do we have for thinking this improvement is exponential? Kurzweil identifies two 'resources' of evolutionary processes that explain why they are exponential. First, he explains, 'Each stage of evolution provides more powerful tools for the next'.[32] A good example is the genetic code. The advent of genetic sentences spelled out using four nucleotide letters, A for adenine, T for thymine, C for cytosine, and G for guanine, was a hugely important evolutionary development. It enabled an acceleration of the evolutionary process by allowing natural selection to choose among the vast profusion of designs for organisms that could be spelled out using different arrangements of a rich heritable alphabet. A further acceleration of the evolution process came with the advent of sexual reproduction. The products of a sexual union combine genetic material from two different organisms. Sex permits a wider variety of genetic combinations than those provided by asexual reproduction. Kurzweil compares these evolutionary innovations with 'the advent of computer assisted design tools' which allow the 'rapid development of the next generation of computers'.[33] In both the evolutionary case and technological case a new design accelerates the process that created it.

Kurzweil proposes that 'the other required resource' for continuing exponential growth is the chaos of the environment in which evolution occurs. 'The chaos provides the variability to permit an evolutionary process to discover more powerful and efficient solutions.' According to Kurzweil, this chaos is the natural equivalent of the 'human ingenuity combined with variable market conditions' that abets technological innovation.[34]

The difference between reflexive and passive improvement

It is wrong for Kurzweil to attribute the exponential improvement of information technology to exponential improvement inherent in the evolutionary process. This is because there is a big gap between any acceleration inherent in the evolutionary process and the acceleration that is a routine feature of information technology. The former may undergo sporadic exponential improvement, but exponential improvement is the norm for information technology. An appeal to a variety of progress that may be, under rare and specific circumstances, exponential cannot explain the fact that another kind of progress is uniformly exponential.

Consider the first of the 'resources' of exponential growth that Kurzweil finds in evolution. Certain evolutionary advances did have the effect of accelerating evolution. As Kurzweil notes, one of these was the invention of the genetic code. Another was the invention of sex. But very many evolutionary advances do not have this property. The advances that tended to accelerate the process of evolutionary improvement were rare. Other evolutionary improvements are merely incremental. They involve no acceleration. The challenge for Kurzweil is to explain why the very rare bursts of exponential evolutionary progress should have a greater impact on technology than the more standard non-exponential forms of evolutionary improvement.

Suppose that a genetic mutation improves the efficiency of a bird's wing. Natural selection may lead the gene to spread throughout the population. But the new mutation does not, in itself, accelerate the evolutionary process. The lesson of evolution by natural selection is that genes linked with small improvements can be subject to positive selection and thereby spread throughout an entire population. The smallest improvement in gliding distance can suffice. There is a stark contrast with the accelerating pattern of improvement that we observe in information technology. An advance in computing power in itself accelerates the process of making more powerful computers. A more powerful integrated circuit boosts the power of the computers that chip designers use to invent even more powerful integrated circuits. When a new advance boosts the quantity of data a computer can store, it can be applied to the design of computers that enable the discovery of even better methods of data storage. More efficient Internet data traffic permits designers of computers to better collaborate in the invention of even more efficient methods of using the Internet to transmit and receive data. What is an exceedingly rare event in the evolutionary process is the norm for improvements of information technology.

In other words, advances in information technology play *reflexive* roles in their own improvement, while advances in wing design are confined to *passive* roles in their own improvement. In a population of winged animals natural selection builds on existing wing designs. Improvement happens by way of mutations to or recombinations of the genes that make wings. These confer advantages and are hence spread by natural selection. But the greater efficiency of the wings does not play any role in arranging advantageous mutations to wing design. Rather, they are passive platforms for potential later improvements. If the evolutionary improvement of wings was an example of reflexive improvement then we should expect to see advantageous mutations arising with greater frequency as wings improve. But this is not the case. Better wings do not make advantageous mutations or recombinations more likely to occur to the genes involved in their own construction.

Contrast this with the evolution of the genetic code and the evolution of sex. A key to understanding the dynamism of the evolutionary process comes from Kurzweil's observation that each is an improvement which 'provides more powerful tools for the next'.[35] A genetic code comprising four nucleotides enables biological designs that cannot be expressed with fewer genetic letters. Sex brings diversity in the genomes of organisms that is unlikely to arise by asexual means. Both are reflexive improvements that speed up the evolutionary process. These reflexive advances are rare, breakthrough events in evolution. As we've seen, such advances are the norm in information technology.

Information technology is reflexively improving. Biological organisms subject to evolution by natural selection are for the most part passively improving. We cannot explain the reflexive improvement of information technologies by pointing to the mostly passive improvements of the evolutionary process. Kurzweil's other evolutionary resource does not add much—the chaos of the environment in which evolution occurs is compatible both with passive and reflexive evolutionary progress. It accompanied and drove the evolution of the genetic code, but it also accompanied and drove the evolution of the wing.

The reflexivity of its improvements distinguishes information technology from other technological domains. A standard feature of technological advances is that they enable further advances. A more accurate musket allows the invention of an even more accurate musket. But here the musket is required only to play a passive role. It serves as a platform for later improvements. Those seeking a better musket do so by modifying an existing design. The musket does not enhance the creativity of designers of firearms. Here we have improvement that follows the passive pattern of the enhanced wing in biological evolution. A more accurate musket may enable its possessors to win more battles. It may permit the invention of even more accurate muskets, but it does not unleash new powers of technological innovation. The more accurate musket is not like more powerful integrated circuits or enhanced Internet bandwidth. Advances in information technology speed up later advances in information technology. Advances in musket technology do not have this effect.[36]

Exponential technological improvement is infectious

The previous observations seem to explain why progress in medicine fails to match progress in information technology. There is improvement in cancer medicine but it is not reflexive. The latest therapy for chronic lymphocytic leukaemia does nothing to accelerate the arrival of even better therapies for the disease. The problems of finding better cancer treatments do not become successively easier with each advance. Deprived of this consistent motor of exponential improvement we should expect to see a somewhat haphazard pattern of progress against cancer. Sometimes it will speed up; sometimes it will slow down. But this speeding up and slowing down is a function of available funding and the intelligence, dedication, and luck of the people who happen, at any given time, to be engaged in the research.

This conclusion is unduly pessimistic. I now offer an argument that we should expect progress in medicine and in a variety of other well-being technologies to increasingly conform to the pattern of exponential improvement characteristic of information technology. This is because of the *infectiousness* of exponential technological improvement.[37] Some biological infections spread by means of a vector. In the case of malaria, a common vector is the *Anopheles* mosquito. In the case of rapid exponential improvement the vector is information technology. Information technology is currently transferring rapid exponential progress to an increasing range of well-being technologies. Once the Plasmodium parasite establishes itself in a host, it multiplies. In an analogous way, once exponential progress establishes itself in a well-being technology, its influence tends to become more powerful. As this influence increases, the pace of improvement in that technology approximates more closely to the rapid exponential pattern characteristic of reflexively improving information technologies.

How might information technology exercise this influence? Consider what the technology commentator Kevin Kelly says about the relationship between information technology and other technologies:

> Moore's Law represents the acceleration of computer technology, which is accelerating everything else. Faster jet engines don't lead to higher corn yields, nor do better lasers lead to faster drug discoveries, but faster computer chips lead to all of these. These days all technology follows computer technology.[38]

Today, information technology is central to the progress of a wide variety of technologies. When information technologies contribute to the progress of a passively improving well-being technology they impart some of their rapid exponential improvement. In effect, they give the development of that technology a push.

There are many examples of this phenomenon of an infectious spread of rapid exponential improvement from information technology. The Airbus A380 jet airliner, which made its first commercial flight in 2007, incorporates a variety of technological advances directly sourced from information technology. Construction did not begin without extensive computer testing of what would be the world's largest commercial jetliner. This testing went far beyond what would have been possible for human aeronautical engineers guided by balsawood models and their intuitions about which designs would fly and which wouldn't. The pilots of the A380 receive assistance from revolutionary networked computers. This is not an aircraft that a joystick-equipped pilot trusts herself to fly entirely by feel. Computers exercised a much less substantial role in the design during the 1960s of the Boeing 747, the former reigning behemoth of commercial air travel. The period of time separating the 1960s design of the 747 and the 2000s design of the A380 saw a considerable displacement of the passively advancing technologies by reflexively advancing information technologies. As passively improving technologies become less influential they act as less of a drag on technological progress.

Information technology is making a variety of increasingly important contributions to cancer medicine. As it does, it tends to accelerate progress. Our understanding of the causes of cancer is now informed by vast databases of information about the genetics and lifestyles of people who get cancer and people who do not. Computers reveal differences in susceptibility to cancer that would be invisible to the unaided human intuition of cancer researchers. People today express surprise that it took so long for doctors to hypothesize a link between lung cancer and what was then viewed as the inherently healthy activity of cigarette smoking. Programs that prowl vast stores of data should reveal many other connections between aspects of intuitively healthy lifestyles and cancer. Surgeons tasked with the removal of tumours can now find their incisions guided by specially programmed computer assistants. Advances in artificial intelligence help in the diagnosis of cancer.

The infectiousness of exponentially improving information technology suggests that progress in cancer medicine will tend to increasingly approximate to the rapid exponential improvement apparent in computers. The greater the contribution of information technology to cancer medicine the more quickly it will accelerate.

What is true of cancer medicine is true of other well-being technologies. Information technology is playing an increasingly important role in the design of our cars. The Google driverless car provides further opportunity for information technology to accelerate automotive technology. By the end of 2013 the Google

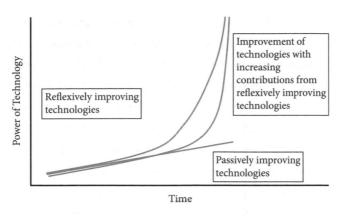

FIG. 2 A well-being technology infected by reflexively improving information technologies.

Chauffeur software was autonomously directing cars along the highways and local roads of a number of US states. As information technologies make increasingly significant contributions we should expect to see a further acceleration of progress in the well-being technology of the automobile.

We can draw some conclusions about the trajectory of a well-being technology towards which information technology makes an increasingly substantial contribution. It will tend increasingly to conform to the pattern of rapid exponential improvement that today characterizes information technology. But so long as passively improving technologies remain, we should not expect to see improvement that quite matches that of information technology (see Fig. 2).

Concluding comments

This chapter has explored the case for the exponential improvement of well-being technologies. I reject Kurzweil's claim that our technologies improve exponentially because we are evolved beings. There are significant differences between the varieties of improvement that tend to result from evolutionary processes and the varieties that are characteristic of exponentially improving technologies. Information technologies undergo reflexive improvement. Evolution, by contrast, is a process of predominantly passive improvement. Presently, many well-being technologies seem to be passively improving. But there is no unbridgeable gulf between the integrated circuit and the technologies that most directly enhance human well-being. The rapid exponential improvement of information technology is infectious. It tends to spread to well-being technologies. Progress in an infected well-being technology tends to accelerate.

The notion that technological progress is the consequence of a law may seem to suggest that we have no meaningful scope for choice about its pace and direction. Some presentations of technological change seem to send this message. The many graphs of exponential progress in Kurzweil's 2005 book *The Singularity is Near* describe the advance of technology in way that makes its acceleration seem both predictable and inevitable. Kurzweil observes that it has been sustained through social arrangements ranging from the despotic and resistant to change to the democratic and dynamic. What we decide, either as individuals or collectively, seems not to make much of a difference. But the notion that our collective choices have little effect on technological progress is mistaken. Exponential technological improvement is a conditional law. It assumes that human beings make certain choices in respect of their technologies—we will try hard to improve them. But the law leaves us free to decide differently. Different choices can make the law of exponential technological improvement less applicable to a technology by causing those responsible for its improvement to less straightforwardly satisfy the law's conditions. There is thus no incompatibility between the claim that technological improvement results from a law and the claim that human inventors, regulators, and users of technology can influence the pace of technological progress. When proponents of rapid technological progress complain of a lack of support, they accept the common-sense idea that different choices by a society's leaders or members can affect the pace of progress.

We can choose to what degree and in what manner we collectively fulfil the assumptions of the law of exponential technological improvement. Radical optimists urge that we do our best to satisfy all the conditions of the law. According to them, we should do what we can to ensure that progress in well-being technologies gets as close as possible to the tempo characteristic of information technologies. I reject this suggestion. In Chapters 3 and 4 I address the question of how technological progress should be expected to boost human well-being.

3

DOES TECHNOLOGICAL
PROGRESS MAKE US HAPPIER?

The excitement that greets the release of new iPhones and the intense speculation about new medical advances are testaments to a strong and widespread belief in the positive effects of technological progress on well-being. We believe that more powerful well-being technologies enhance human well-being. Radical optimists present what they take to be an implication of this belief. Exponentially improving well-being technologies should produce enhancements of well-being of corresponding magnitude.

In this chapter I consider two obstacles to the translation of improvements in well-being technologies into enhancements of subjective well-being. The first arises in respect of an observed tendency for technological progress to make people sadder, not happier. There appears to be something fundamentally self-defeating about enhancing happiness by technological means. The second obstacle arises in respect of human subjective well-being itself. There is strong evidence supporting the existence of hedonic baselines set by our genes and early experiences. The phenomenon of hedonic adaptation tends, through life's ups and downs, to return humans to this set point. It seems to prevent positive events, including the acquisition of more powerful well-being technologies, from having anything more than temporary effects on subjective well-being.

Neither of these obstacles falsifies the notion that technological progress can significantly enhance well-being. Perhaps we are sometimes fooled by those marketing new well-being technologies into overestimating their benefits. But the contributions of more powerful well-being technologies to individual subjective well-being are real. Chapter 4 demonstrates that these enhancements of individual subjective well-being fail to translate into corresponding improvements for the societies to which these individuals belong.

The traditional paradox of progress

This chapter and Chapter 4 describe a 'new' paradox of progress. I call it 'new' to distinguish it from a widely discussed 'traditional' paradox of progress.

The new paradox concerns the translation of improvements of individual subjective well-being to improvements for the societies to which those individuals belong. The traditional paradox of progress, in contrast, directly addresses individual benefits. It casts doubt on the idea that individuals can benefit from the invention of more powerful technologies.

This traditional paradox emerges from the observation that improvements of wealth or technology that ought to improve human well-being seem not to. In a recent statement of the idea, Richard Wilkinson and Kate Pickett say:

> It is a remarkable paradox that, at the pinnacle of human technical and material achievement, we find ourselves anxiety-ridden, prone to depression, worried about how others see us, unsure of our friendships, driven to consume and with little or no community life.[39]

Inhabitants of the affluent world today are significantly wealthier than were their forebears. They have access to medical therapies unavailable to earlier generations. They benefit from a variety of labour- and time-saving devices. In spite of this manifest progress, people don't seem happier. Rather than straightforwardly improving health, better medical therapies seem to have produced a large number of 'worried well'. Material progress seems to have resulted in status anxiety and hence to have reduced happiness rather than enhancing it. It has weakened rather than strengthened social bonds. We are befuddled by modern life's complexities and bewildered by the array of choices that confront us.

One recent study of the effects of social networking technologies on happiness seems to support this scepticism. It appears to demonstrate an effect on well-being the opposite of that predicted by the technology's marketers. A simple model of how social networking technologies work should make it obvious that they will make us happier. Social isolation is a recognized cause of human unhappiness. Since Facebook enhances the capacity to connect socially it should straightforwardly enhance our happiness. You no longer have to leave your home to be richly socially connected. It seems, however, that human beings are more complex than assumed by a simple model according to which an increase in the number of social connections boosts happiness. Ethan Kross, a psychologist at the University of Michigan, discovered a tendency for Facebook to increase feelings of sadness and loneliness.[40] There was an observed positive correlation between feelings of unhappiness and isolation and the amount of time on Facebook. One explanation for this tendency is the ease with which Facebook permits comparisons between our achievements and those of others. When on Facebook, we tend to seek out people who resemble us in ways that we care about. But the people with whom we interact are inclined to be selective in what they say about themselves. In face-to-face interactions we can often see through this positive bias. However, the cues that we tend to pick up on in face-to-face conversations are less available on social networking sites. And so we come away from Facebook with the impression that many, perhaps most of those to whom we compare ourselves, are actually doing better than us. Our feelings of inadequacy are amplified. A technology that was supposed to counteract the effects of social isolation on well-being may, in some circumstances, have the opposite effect.

The most forthright presentations of these conclusions seem difficult to accept. Is Facebook guilty of perpetrating a massive fraud on its customers? Do many of the technologies that people spend considerable sums of money on in the hope that they will make them happier actually have the reverse effect? Surely this cannot be right. A less alarmist conclusion is warranted. The psychological and emotional mechanisms that produce different levels of subjective well-being are more complex than some designers of well-being technologies suppose. Facebook can connect you with people in ways that make you happier. However, if you find it displacing the more traditional forms of interaction that occur when you meet friends in a café or pub then you may find that it begins to have the reverse effect. The problem is not so much with new well-being technologies themselves, but rather with certain ways of using them.

I think that there is a connection between the traditional paradox of progress and the 'fundamental paradox of hedonism' described by the nineteenth-century utilitarian philosopher Henry Sidgwick. Sidgwick describes the fundamental paradox of hedonism as the idea 'that the impulse towards pleasure, if too predominant, defeats its own aim'.[41] The direct and deliberate pursuit of pleasure by individuals seems not to produce pleasure. An overriding purpose of celebrity television channels such as *E!* seems to be to confirm Sidgwick's observation. Its many stories about the comeuppances of Lindsay Lohan and Charlie Sheen demonstrate that those who should have the best opportunity to achieve pleasure and dedicate themselves to its pursuit do not achieve it.

The paradox of hedonism seems to make pleasure different from goals that do reward direct commitment. One attempts to gain physical fitness by committing oneself to an exercise programme. The better one's dedication to and compliance with the programme, the more impressive should be its results. People who exercise for an hour a day should tend to see sharper increases in their physical fitness than do people who exercise for a single hour a week. A policy of directly pursuing happiness seems different. It does not produce results commensurate with the diligence with which it is pursued.

The awkward relationship between achieving a goal and deliberately pursuing it described by Sidgwick is actually not unique to the pursuit of pleasure. Here's a strategy that you may choose to implement should you find yourself playing golf against someone whose game is manifestly superior to your own. You should direct your opponent to try her hardest to play her best game of golf. More specifically, you should direct her to produce these best golf strokes by a process of consciously scrutinizing every component of her game. She should focus on every detail of her swing, play close attention to the nature of the contact of her club with the ball, and consciously and deliberately supervise her follow-through. In short, she should seek to consciously optimize every aspect of her golf game. This close conscious supervision is very likely to produce golf strokes inferior to those she would have produced had she resolved to follow Nike's advice and 'just do it'. Strategies of conscious deliberation are useful when one is seeking to acquire a new skill. Someone who is learning how to play golf should consciously seek to implement a teacher's instructions about every aspect of his shot. But what works well for beginners does not work well for experienced golfers. The latter have an automated access to programmed sequences of movements informed by years of practice and play. When your opponent reverts to the strategies best employed by beginners, her access to these results of years of training is disrupted.[42]

The traditional paradox of hedonism does not show that pleasure cannot be pursued. Rather, it cautions against certain ways of pursuing it. Improvements of happiness are, in general, best pursued indirectly. What is true for experienced golfers should also be true for people with sufficient life experience to separate activities that lead to well-being from those that do not. It is not surprising that the happiest people are not the libertines consumed by a self-conscious pursuit of pleasure. They are those who commit to worthy goals for their own sake. Happiness seems to emerge as a by-product of this commitment. If you want to be happy, find some goal that is meaningful to you. Sometimes conscious intervention may be required if you find that you have fallen into bad habits of thought. Perhaps you have internalized a negative view about yourself that has become an obstacle to the pursuit of happiness. This would be analogous to the experienced golfer who reapplies the learning strategies of beginners in respect of a part of her game that is misfiring. But happiness does not generally result from introspection on the minutiae of the plans that guide our actions. Happy people are more outwardly focused, seeking goals that are worth achieving in themselves.

This lesson about how improvements of well-being are best pursued applies to the design of technologies. Well-being technologies whose design too directly targets happiness may not have the intended effect. Internet pornography, violent computer games, and the innovation that fused the doughnut and the croissant to form the cronut seem designed to directly target the brain's pleasure centres. They may, as a consequence, improve subjective well-being less effectively than do technologies that promote activities that we find more meaningful. Those who make persistent use of technologies directed at the production of pleasure may notice an influence on their thought patterns. They find themselves opting for cheap thrills. But this message is no problem for well-being technologies less directly targeted at the production of pleasure. The sleek design and ringtones of a new smartphone directly stimulate our senses. But the new smartphone enhances other activities that we are more likely to find more valuable. It enables new forms of social interaction. It permits Google searches for information necessary to implement a meaningful life plan. Advances in medical technology lengthen lives and alleviate the effects of chronic illness, granting access to a wide range of valuable experiences.

The traditional paradox of progress cautions against certain ways of using technology to achieve happiness, but it is certainly no barrier to the enhancement of well-being by technology.

How we hedonically adapt to new well-being technologies

One of the most important findings of recent empirical research on well-being is the existence of hedonic set points. These set points seem to emerge from a complex interaction between genes and early experiences.[43] The triumphs and tragedies of human lives do boost or depress subjective well-being. However, people tend, over time, to return to their set points. Winners of lotteries experience a significant boost in well-being immediately after learning of the win. One year after the win, their levels of subjective well-being have returned to their hedonic set points. They are approximately as happy one year after the win as they would have been had their ticket not been drawn.[44] Something similar seems to be the case in respect of negative events. People who become significantly disabled as the result of an accident experience an abrupt dip in subjective well-being immediately after the accident. According to some scientists of subjective well-being, one year after the accident they have bounced back to levels of subjective well-being at or near to those that they would have experienced had the disabling accident not occurred.

A metaphor prominent in discussion of the propensity of adaptation to erode gains in subjective well-being is the hedonic treadmill, introduced to the psychological literature in 1971 by Philip Brickman and Donald Campbell.[45] Someone who finds herself on a treadmill must go forward just to remain stationary. However, treadmills permit no significant forward motion relative to the ground upon which the contraption sits. Gym-goers who insist on running faster than is appropriate for the setting of the treadmill get injured.

This treadmill is not a perfect metaphor for the effects of hedonic set points. To describe the propensity of adaptation to respond to setbacks we would have to imagine a treadmill that runs in reverse. More importantly for this book's purposes, it understates the real hedonic gains from good events in one's life. When you are exercising on a treadmill, accelerations in your pace should exactly match accelerations of the treadmill. You have only the most limited scope to run faster. Research on hedonic adaptation suggests that real, albeit temporary boosts in subjective well-being do occur. Some gains, though temporary, may nevertheless be long-lasting. In the context of an eighty-year lifespan, six months of joy from a lottery win could represent a significant net improvement in happiness.

Consider how we might apply conclusions about hedonic adaptation to technological progress. The radical optimists present a picture of technological progress according to which we receive a wide variety of improved well-being technologies that boost subjective well-being. If we suppose that these hedonic boosts are temporary then they will, over time, erode completely. This would not prevent an accelerating pace of technological progress from delivering improved well-being technologies that collectively produce significant long-lasting improvements of subjective well-being. As soon as you are beginning to tire of your new iPhone, Apple has a new iPod ready for your consideration. When that gets old, there's a new iPad. Essentially, all Apple has to do is to keep the iGadgets coming. Ideally the arrival of a new well-being technology would be timed for when the hedonic benefit of the last innovation had worn off. The danger here might be from a kind of meta-adaptation in which one ceases to derive any hedonic benefit from new examples of certain species of well-being technologies.

Those in the grip of meta-adaptation might be tempted to para-phrase the line in Alan Bennett's play *The History Boys* about history being "just one fucking thing after another." They might view Apple as producing just one device after another.

Hedonic adaptation is adaptive in evolutionary terms. Sonja Lyubomirsky puts it well when she says:

> If people's emotional reactions did not weaken with time, they would not be able to discriminate between more significant stimuli (i.e., new events that offer new information) and less significant stimuli (i.e., past events that should fade into the background).[46]

Applied to the acquisition of new well-being technologies, it is apparent that if the feeling of bliss that you felt upon acquisition of a new Samsung Galaxy did not fairly swiftly erode you would not be motivated to pursue other goals. You could lead a fully satisfied life by means of serial re-acquaintances with your smartphone. The idea of smartphone owners perpetually blissed out by their devices may conform to a popular stereotype, but it does not describe the behaviour of most of them. This pattern of behaviour would not be biologically adaptive.

Complete or incomplete hedonic adaptation?

One debate about hedonic adaptation concerns its degree. Accord-ing to one account of the degree of hedonic adaptation to positive or negative life events, the attractive power of an individual's hedonic set point is sufficient to restore him to exactly the level of subjective well-being that he would have experienced without the significant life event. Any good or bad hedonic effects will be temporary. After these blips or dips you end up with a level of

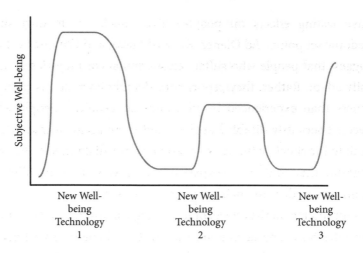

FIG. 3 An individual's complete hedonic adaptation to a series of improved well-being technologies.

subjective well-being no different from what you would have experienced had the good or bad thing never occurred (see Fig. 3). The suggestion that we completely adapt to improvements can be applied to improvements of well-being technologies. An improved asthma inhaler or MP3 player produces temporary improvements to which one swiftly adapts.

This view seems to significantly undermine the motivation of actively pursuing good things and actively avoiding bad things. For example, it seems to foster an attitude in which one should calmly ride out misfortune, awaiting the return to one's hedonic set point. If it's going to be difficult to avoid a future disaster then why bother? You can be confident that its ill effects are only temporary.

Current thinking suggests that set points do not exercise quite this degree of power over our hedonic states. Although there is significant hedonic adaptation, great successes and tragic setbacks have lasting effects on people's lives. Such events may shift hedonic set points. Ed Diener, Richard Lucas, and Christie Scollon suggest that people who suffer serious spinal cord injuries do not fully adapt. Rather, they report considerably lower levels of happiness than experienced by non-disabled people.[47] People who become seriously disabled as the result of an accident may adapt both to the shock of the accident and its immediate effects. But one year out, they continue to experience the effects of the disability. It continues to depress their subjective well-being.

Our interest in the effects of technological progress directs us to consider how long-lasting are the positive changes in well-being that result from the acquisition of a more powerful well-being technology. As Lyubomirsky notes, there is less research on longer-term hedonic responses to positive changes than there is on responses to negative changes.[48]

It's clear that we do, to a considerable extent, adapt to technological advances. A more capable smartphone typically brings more pleasure when it is new and its many features and functions feel fresher than when it is one year old and your friends' phones are cooler. The boost in subjective well-being brought by the newness of the smartphone erodes. But it is unduly pessimistic to suppose that this erosion must be complete. Hedonic adaptation occurs, but it need not be total. As will become clear, a new well-being technology may leave a lasting improvement of subjective well-being.

The following discussion follows a model of hedonic adaptation proposed by Lyubomirsky and her colleague Ken Sheldon. This is the Hedonic Adaptation to Positive and Negative Experiences (HAPNE) model. In Lyubomirsky's presentation of HAPNE, hedonic adaptation occurs by two distinct psychological paths.

One of these paths offers the possibility of adaptation-thwarting interventions. Such interventions should be spurned in respect of negative events but actively sought in respect of positive events. Lyubomirsky says 'The first path specifies that the stream of positive or negative emotions resulting from the life change (e.g., joy or sadness) may lessen over time, reverting people's happiness levels back to their baseline'.[49] This path directs adaptation to a new well-being technology by means of a reduction in the positive emotions that it tends to produce. We tend to completely adapt to the non-functional aspects of a Samsung Galaxy smartphone such as its distinctive aesthetic and cute ring tones. Several months into their ownership of a Samsung Galaxy, few people are busy cycling through its ring tones or gazing admiringly at the clean lines of its design. They have hedonically adapted to these features and no longer experience much pleasure from them.

Lyubomirsky proposes a 'second, more counterintuitive path' that promises more lasting returns from positive life events. She notes that adaptation seems fastest in response to 'constant stimuli'. Adaptation can be thwarted or at least delayed when one partakes in activities 'which entail persistent effort and engagement in a self-directed process. Such activities have the property that they can be varied and episodic and can produce a fluid and diverse set of positive experiences, opportunities and possibilities'.[50] This phenomenon might explain the observed pattern of incomplete adaptation to a serious spinal injury. Those who suffer such injuries do not have a single, invariant unpleasant experience. If they did, they would adapt. Rather they find that their injury limits what they do in a range of unexpected ways.

One swiftly adapts to a new Samsung Galaxy's ring tones. But the smartphone gives rise to other more enduring pleasures. It enables access to social networking sites. It enhances your ability to be productive while travelling. These aspects of a smartphone produce positive experiences that are often novel and surprising. They demand active engagement on the part of smartphone users. Unlike a meowing kitten ring tone, these are not things that users of a smartphone passively enjoy. Lyubomirsky's HAPNE model of hedonic adaptation suggests that these enhancements of well-being are more persistent. When new well-being technologies produce good effects that are relatively constant and predictable, our hedonic adaptation will tend to be complete. When a new technology enables dynamic engagement with it and requires significant effort and leads to variable and unpredictable positive experiences, then our hedonic adaptation will be significantly retarded and when it occurs may be incomplete. It will tend to leave a long-lasting hedonic residue (see Fig. 4).

Many well-being technologies produce the second variety of effect. One acquires the variable and surprising pleasures of posting frequent tweets by means of a technology that enables you to log into Twitter. In other cases a new well-being technology stands in a more distant but nevertheless real causal relationship with the variable and surprising pleasures it permits. A well-being technology that extends lifespans permits many novel and unexpected pleasures. Those who live longer will have longer to hedonically adapt to new well-being technologies, but a combination of the lengthening of their lives and the accelerating pace of technological progress means that they will experience many more technological novelties.

It's when we combine this conclusion with the conclusions of Chapter 2 that we recognize the scale of possible improvements of subjective well-being from technological progress. According to

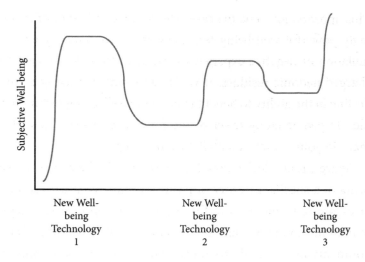

FIG. 4 An individual's incomplete hedonic adaptation to a series of improved well-being technologies.

the radical optimists, technologies are becoming more powerful at an ever-increasing pace. If Lyubomirsky's model serves as an adequate account of the propensity of new enhancement technologies to boost subjective well-being, then we should expect adaptation-proof hedonic residues to accumulate at an accelerating pace. We can measure the net long-term improvement of subjective well-being from technological progress by summing the value of the hedonic residues produced by more powerful well-being technologies. We should, over time, experience an accumulation of hedonic residues that will make our lives significantly better.

If life were a simple accumulation of positive hedonic residues from technological progress then we should expect to see increasing levels of happiness as people age. Retirement homes would be filled with people enjoying a lifetime's worth of hedonic savings.

This appears not to be the case. The positive influence of increasingly powerful well-being technologies is countered by an accumulation of negative experiences. Many of these leave negatively charged hedonic residues. It is also apparent that aging causes a decline in the ability to benefit from many well-being technologies. Elderly people enjoy fewer of the features of a new smartphone than do young, more mentally flexible people.

Figure 4 is an idealization. It represents only the effects on well-being of a specific category of positive event—your acquisition of or exposure to more powerful well-being technologies. A typical human life contains many such events. But it also contains abandonments by romantic partners, dismissals from jobs, and root canal dental procedures. We would present human lives as unrealistically miserable if we restricted our focus to those negative events. But we also get an unrealistically upbeat presentation if we restrict our focus to the introduction of new smartphones and innovative therapies for coronary artery disease.

We should not expect the acquisition of new well-being technologies to feature among the hedonic highlights of a typical human life. The feeling of requited romantic love should, for most people, be valued much more highly than the acquisition of a new Sony PlayStation. The highest hedonic highs of human lives tend to result from events towards which technology makes only a peripheral contribution. It's likely that the joy of becoming a parent for the first time was as keenly felt 2,000 years ago as it is today.

But these points notwithstanding, there should be a fairly constant upward influence through all of life's ups and downs from the discovery of more powerful well-being technologies. If these technologies are becoming exponentially more powerful then such improvements should become increasingly apparent. This should be so even if technological progress has a less significant effect on well-being than do other influences.

Consider an analogy. There is now not much doubt that human activities are influencing the global climate. People are increasingly inclined to attribute especially violent storms to anthropogenic climate change. But this assumes too simplistic a view of the effects of human-caused climate change on our weather. One of the great difficulties in recognizing the effects of human influences comes from the background variability in climatic conditions. Significant fluctuations in weather patterns occur quite apart from any human influences. Violent storms predate anthropogenic climate change. For any given storm in the second decade of the third millennium of the Common Era, the human contribution may be quite minor. But a series of small contributions can add up to a significant net effect on the climate considered as a whole. An increase in greenhouse gases may begin to produce effects of such magnitude that they can clearly be distinguished from normal background variation in weather patterns.

A similar relationship could obtain between improvements of well-being technologies and well-being. A human life comprises many good and bad events. Better access to well-being technologies is likely to have only a small impact on any one of these events. The event of being dumped by your spouse is likely to make a big difference to your well-being. The fundamentals of the relationship have little to do with the technologies that you and your spouse possess. But improved well-being technologies may nevertheless exercise a significant influence on how a typical human life unfolds. Perhaps exponential improvement of well-being technologies will soon produce hedonic effects clearly distinguishable from life's expected ebbs and flows.

Concluding comments

This chapter focused on the propensity for technological progress to improve individual well-being. Two obstacles seem to block benefits from more powerful well-being technologies. The first obstacle results from a purported tendency for technological progress to reduce well-being by increasing anxiety and concern about our social standing. Such obstacles to improved well-being seem to emerge from the way in which we pursue improvements of well-being. Too deliberate a focus on them threatens to make things worse rather than better. I interpret this as a caution about how best to pursue technological enhancements of happiness. They are often best sought indirectly. The next obstacle arises in respect of the confirmed phenomenon of hedonic adaptation. There is strong support for the existence of hedonic set points to which we return after a positive event, such as the acquisition of a new well-being technology. It is likely that some of the hedonic benefits of technological progress are temporary, but some should be longer lasting. A steady accumulation of these hedonic residues can have a powerful impact on an individual's life.

The scene is now set for the description of a new paradox of progress: many of these individual benefits of progress fail to translate into improvements for the societies to which individuals belong.

4

THE NEW PARADOX
OF PROGRESS

In his monumental work *The History of the Decline and Fall of the Roman Empire,* Edward Gibbon pronounced the Antonine dynasty of ancient Rome, a period that commenced towards the end of the first century CE and ran for much of the second century, to be 'the period in the history of the world during which the condition of the human race was most happy and prosperous'.[51] This was a period of Roman history favoured by mainly competent and non-psychopathic emperors, including Trajan, Hadrian, and Marcus Aurelius. It was also a time of significant industrial expansion. The writing of Gibbon's history spanned the years 1776 to 1788 so we can suppose that he meant to compare ancient Roman happiness with the happiness of people from historical eras ranging from the first century up until the happiness of people of his own Georgian England.

Suppose that we exclude Gibbon's claim about the relative prosperity of Antonine Rome and eighteenth-century England. Our early twenty-first-century understanding of prosperity tends to be narrowly economic. It's hard to think of a measure of economic development that would have placed the Roman Empire of 100 CE ahead of eighteenth-century England. This was an England experiencing the foreshocks of the Industrial Revolution. What of the claim that the people of Antonine Rome were happier than the eighteenth-century English?

In this book I interpret claims about happiness as claims about subjective well-being. Historians may dispute this translation of Gibbon's word 'happy'. There is some reason to think that the word 'happy' meant something different for Gibbon from what it means to us. The identification of happiness with something subjective became increasingly predominant in the later part of the twentieth century. Before then, happiness was more or less synonymous with the observable phenomenon of success. We treat the question 'Is Barack Obama happy?' as a meaningful inquiry about the subjective experiences of the forty-fourth president of the United States. For Gibbon, Obama's happiness might have been something close to a necessary truth. Unhappy presidents would, for him, be a bit like married bachelors.

Here I interpret Gibbon as asserting that the Romans would have given higher estimates of life satisfaction and reported superior ratios of positive to negative experiences. I choose to read him as making a claim about the subjective states of first-century Romans because, as we shall see, it comes close to being true.

Gibbon's point seems quite specific to happiness. He is unlikely to have made analogous claims about other indices of technological progress. Advances in military technology would have seen even the best-disciplined Roman legion of 100 CE go down to defeat under the fire of the cannons and muskets of a European army of 1776. The farmers of Europe's eighteenth century were dramatically increasing crop yields by means of agricultural mechanization. The lesser yields of Roman farms could not be maintained without slave labour. Scepticism about the positive effects of technological progress seems warranted specifically in respect of well-being.

Gibbon versus Ridley on historical happiness

Could Gibbon's claim be true? One modern writer who almost certainly would disagree with this assessment of the effects of progress on well-being is the radical optimist Matt Ridley. In his 2010 book, *The Rational Optimist*, Ridley compares the Europe and North America of 1800, just after Gibbon's time of writing, with the Europe and North America of our own day. Ridley bemoans the common tendency to romanticize the past. Many people view times gone by as times of 'simplicity, tranquillity, sociability, and spirituality'.[52] Ridley undertakes to disabuse us by presenting some home truths about home life in the past. He invites us to imagine a family living in 1800 somewhere in Western Europe or North America. Ridley allows that our minds reach most naturally to a depiction of the family gathered round the hearth in a simple house as father reads aloud from the Bible, mother cooks up a beef and onion stew, and a baby is comforted. All about are expressions of bucolic bliss—water is poured into earthenware mugs, horses are fed, and birds sing. There are no drug dealers or dioxins. This, according to Ridley, significantly distorts the past. Ridley offers a selection of details that the idealized narrative overlooks. Father has pneumonia. The baby needs comforting because he has smallpox. The sister awaits a future as a 'chattel of a drunken husband'.[53] The mother must endure constant toothache. The stew is, of course, disgusting.

Ridley proposes that we should be exceedingly grateful for the progress—moral, economic, and technological—that separates the family of 1800 from the families of today's Europe and North America.

It's possible that Gibbon is the victim of distorting idealism about the past. He seems attracted to the members of Rome's ruling class whom he prefers to imagine stoically resisting the forces of moral degeneration that would eventually result in the empire's decline and fall. I will have something to say about Gibbon's selective vision of the past. But for now I focus on a bias that has entered into Ridley's thinking. This bias leads Ridley to significantly overstate how bad it was to be alive in 1800. I suspect that when the effects of both sources of bias are compared, Gibbon emerges as closer to the truth than Ridley.

The perils of attitudinal time travel

When we attempt to imagine life in 1800 we tend to misrepresent what it would be like to be alive then. We inadvertently send back in time some of our attitudes and beliefs. We engage in attitudinal time travel. A marooned time traveller from the early twenty-first century forced to live in conditions similar to those described by Ridley would find them terrible. She has knowledge and experience of the moral, economic, and technological circumstances of our time. Her misery is, in large part, a consequence of a comparison of the past with the present. It's difficult to avoid introducing this knowledge into our attempts to imagine what life was like for people in the past.

Attitudinal time travel is a standard feature of historical fiction. It occurs when an author awards to the characters in a novel set in the long European summer of 1913 an inchoate sense of foreboding. This suits her story-telling purposes even if she knows they cannot have predicted the carnage of World War One. We can detect attitudinal time travel in characters' interior monologues. Suppose that our hero is a magistrate living in the Rome of 100 CE. He is likely to be described as thinking about things predictably of interest to us, but which will have been utterly unremarkable to the Romans of 100 CE. He notices and reflects on the extreme bad breath of the people he speaks to, standard features of an age that lacked dental hygienists. He is appalled by the human faecal matter that occasionally drops on to the road from upper storey windows. When on the streets of ancient Rome, shit often happened; the Romans had many centuries to wait for proper flush toilets. The interests of readers from the early twenty-first century inform the magistrate's interior monologue. We might compare this with a novel written by a twenty-third century author but set in the early twenty-first century whose central character takes a sustained interest in the fact that the roads are filled with four-wheeled personal transport vehicles that emit toxic fumes. For authentic denizens of the early twenty-first century, they're just cars.

High on the list of time travelling attitudes in historical fiction are our moral beliefs. A well-researched movie set in the London of 1700 may get many of the physical details of life at that time correct. But its central protagonists probably do not express some of the moral beliefs that would have been prevalent at that time. They are unlikely to be presented agreeing with the notion that members of the darker races are natural slaves or as believers in the notion that it is fitting and proper that a wife receive daily beatings. Here, attitudinal time travel occurs for the simple reason that it would be impossible for such characters to engage the sympathies of modern moviegoers. Such attitudes are reserved for the movie's villains.

Some degree of attitudinal time travel may be essential in historical fiction. However it prevents us from understanding how good or bad life was in the past. It is near impossible for people alive today to imaginatively forget what they know about the moral, economic, and technological conditions of the early twenty-first century. They cannot avoid comparisons based on knowledge that was unavailable to people in the past. Ridley's account of the miseries of 1800 relies on attitudinal time travel. When we imagine what it was like to live then, we find it very difficult to forget what we know about life today.

To see the error in Ridley's way of thinking about the quality of past lives, consider how future people will think about us. If the surveys are to be believed, most people today have quite high levels of subjective well-being.[54] Attitudinal time travel has the potential to dislodge this assessment. Let's assume that humanity avoids any of the apocalyptic scenarios that feature in Hollywood movies, and that there is an increasingly exponential rate of improvement as information technologies continue to infect well-being technologies. What judgments would future people make about how good it is to be us? If they are not careful to avoid attitudinal time travel, they will take the same attitude towards our lives that Ridley encourages us to take of the lives of people in the early 1800s. They will be appalled by our diseases and disgusted by our food. Yet the fact that a resident of the twenty-third century would not trade places with us does not mean that we should revise downward our assessments of the quality of our own lives. It means that we should resist assessments of the quality of our lives from an entirely alien perspective.

We know enough to avoid a cross-cultural equivalent of attitudinal time travel. People who loathe raw fish tend not to attribute their feelings of disgust to Japanese people eating a plate of sashimi. Such feelings would render inexplicable the expressions of contentment on the faces of the diners. We don't have such direct access to the expressions of suffering or happiness of the people of the rich world of 1800 or of Antonine Rome to correct our beliefs about how good or bad they found their lives. But that doesn't mean that we are clueless. We can appeal to an evolutionary hypothesis about the psychological and emotional states that produce well-being.

Hedonic normalization

According to the received view among anthropologists, human beings are unchanged in our biological fundamentals since the Pleistocene, a period that commenced approximately 2.6 million years ago and finished around 12,000 years ago. During this time humans were mostly confined to Africa. Those with an interest in human evolutionary history label the environments inhabited by humans during the Pleistocene the environment of evolutionary adaptedness or EEA. Evolutionary psychologists expect that any plausible hypothesis about a human psychological mechanism should be compatible with known facts about the EEA.[55] If there's reason to doubt that a hypothesized significant psychological mechanism could have evolved in the EEA then there's reason to doubt its existence. For example, a tendency to form religious beliefs is a recurrent and significant feature of human minds. These puzzle evolutionary theorists. They seem to impose quite significant costs on believers—the sacrifices required by religion—

in exchange for benefits that seem predominantly spiritual, i.e. non-reproductive. Evolutionary psychologists have invested much effort in an attempt to show how religious beliefs might have promoted survival in a Pleistocene environment.[56]

Consider the implications of the EEA for the psychological and emotional states responsible for producing subjective well-being. The capacity for variable levels of subjective well-being is a significant feature of human minds—it is unlikely to have been a mere by-product of selection for other, more significant mental abilities. Subjective well-being has a biological function. The capacity for high subjective well-being evolved to prompt us to pursue biologically significant ends and to give us hedonic rewards when these ends have been achieved. The biological function of low subjective well-being is to motivate us to avoid outcomes bad for our prospects of surviving and reproducing. When we find ourselves in such circumstances, low subjective well-being motivates us to take positive steps.

Evolutionary psychologists tend to single out the sparsely treed grassland ecosystem of the African savannah for special attention. This was the environment occupied by many humans during the Pleistocene. Perhaps this was the environment that exercised most influence over the evolution of human minds—but humans probably evolved to respond well to other kinds of African environments too. Humans are likely to have spread from the savannah to the coasts and to relatively densely forested regions. Some of these communities may have had lifestyles significantly different from savannah dwellers. This environmental variability suggests a feature of the psychological and emotional mechanisms of subjective well-being. I call this hedonic normalization.

Hedonic normalization: a tendency to form goals and experience hedonic rewards appropriate to the environments experienced as an individual comes to maturity.

Hedonic normalization can be viewed as the intergenerational equivalent of hedonic adaptation. Hedonic adaptation adjusts an individual's well-being so as to enable her to better confront new challenges. If the feeling of bliss resulting from a significant success did not soon fade she would be unmotivated to seek new successes. Hedonic normalization aligns our subjective well-being with our objective circumstances. The hedonic baselines of new individuals are fixed by the environments encountered as they come to maturity. A new technology may significantly increase the well-being of the members of the generation that discovers it. If their children commenced life with the same elevated feelings of well-being, they would be insufficiently motivated to seek new improvements. Hedonic normalization permits the technology to contribute to the hedonic baselines of their children.

The capacity to respond appropriately to a wide range of environments is likely to be an indispensable feature of our evolved mechanisms for generating feelings of high or low subjective well-being. The presence of a mechanism that normalizes human psychology to a wide range of environments distinguishes us from many non-human animals. Naked mole rat psychology evolved to enable naked mole rats to perform behaviours that enhance biological fitness in a very specific environment—systems of underground burrows in desert regions of East Africa. It should not be at all surprising that its inflexible psychology would prevent the naked mole rat from performing biologically adaptive actions when transported out of its desert environment into an arctic environment. Human psychology is different. It evolved to enable us to perform fitness-enhancing behaviour in a wide variety of environments. Biologically salient facts about those environments influence our hedonic set points.

The psychological and emotional mechanisms of subjective well-being would have responded differently to different historical environments. Suppose that you have grown up in the Late Pleistocene. Returning home with sufficient food to feed your family is a significant achievement. Your knowledge of your environment leads you to treat it as such. You know that food is scarce. You know that hunting or gathering can be dangerous. You receive hedonic rewards appropriate to this knowledge. The same evolved propensity grants a lesser hedonic reward to a resident of early twenty-first-century Manhattan who returns home with a heat-and-eat meal picked up at a local supermarket. Bigger hedonic rewards are reserved for achievements that your knowledge of the conditions prevalent in early twenty-first-century affluent liberal democracies permits you to view as significant.

Consider how hedonic normalization affects the experience of dentistry. The dental procedures of Gibbon's day were performed without anaesthetic. They might have involved the fitting of a replacement tooth fashioned out of wood. Doubtless dental patients of that time did not relish the prospect of wisdom tooth extractions. But the process of hedonic normalization does something to reduce dental suffering. Those who grew up in Gibbon's England would have heard accounts of dentistry. They would have accepted dental suffering as an unpleasant fact of life. Accepting it as normal does not, of course, mean that they would not change it if they could. But they believe that they can't.

The experience of dentistry of Georgian England would be significantly worse for a marooned time traveller from our time.

The marooned time traveller has expectations about dental procedures shaped by experiences of late twentieth and early twenty-first-century dentistry. We can make the experience of Georgian dentistry still worse if we imagine its subject to be a time traveller from the twenty-third century whose dental expectations are shaped by miniature robotic products of nanotechnology that, once introduced into a human mouth, instantly and painlessly repair any defect.

Consider now the likely experiences of those who travel forward in time for their dental procedures. A time traveller from Georgian England would doubtless experience great joy at the prospect of receiving early twenty-first-century dental care. People in a possible twenty-third century raised to expect that any dental issue will be perfectly addressed by the introduction of dental nanobots view this as normal. They may experience a wistful relief upon hearing accounts of times when dentistry involved injections and drilling, but they do not feel the happiness I would feel upon learning that my dentist was able to treat me in this way. These different experiences of dentistry are consequences of hedonic normalization.

Note that the mere knowledge that times have been worse and will probably get better does not suffice to cancel or overwrite hedonic normalization. The next time you go to the dentist for what you anticipate will be a fairly involved procedure, attempt one of the following experiments in imagination. Think about what it would be like for a dental patient of 1800 whose teeth need attention similar to yours. Think about the fact that your 'dental professional' could be a barber doing some tooth-extraction on the side. No anaesthetic would be involved. Alternatively, think about what it would be like to be someone in 2300 preparing to gurgle with a solution containing dental nanobots. I doubt these thought experiments would have a very great effect on your experience of dentistry. They would not suffice to overwrite your hedonic normalization to the dental technologies of the early twenty-first century.

When we imagine life in first-century Rome or nineteenth-century Europe or North America we tend to overstate the progress in subjective well-being that has occurred between those times and our own. When we imagine the conditions in which ancient Romans lived, we tend to imagine what it would be like for us to live in that time. To get an accurate picture of how good that life would be, we have to do something that is very difficult indeed. We have to imagine away our knowledge of the well-being technologies of our time. Had I been born in Rome in the first century, I would have no knowledge of more advanced well-being technologies. I would not find the techniques of Roman dentistry a combination of the terrifying and disgusting. I would be hedonically normalized to the conditions of ancient Rome—including its dental procedures. Omitting to take into account the effects of hedonic normalization leads us to significantly overstate the effects of technological progress on subjective well-being.

Technological change is not the only thing that separates the present day from bygone times. There have been changes in social arrangements. Slave-holding societies have been replaced by societies that emphatically reject slavery. We think of these changes as having significantly boosted well-being. Do the preceding claims about hedonic normalization undercut our sense of moral achievement? I think not. What is true is that one can, to some extent, become hedonically normalized to slavery. Children raised as members of servant classes doubtless came to accept these circumstances as normal. This did not prevent them from experiencing levels of subjective well-being lower than those of their owners. Furthermore, the fact that one can be somewhat hedonically normalized to injustice does not render it just. The slaves had a moral entitlement to freedom. The relatively happy among them had the same basic moral entitlement as the miserable. No similar moral complaint is available in respect of an age's primitive well-being technologies. Much as I would like them, I have no moral entitlement to twenty-third-century dental nanobots. It's simply not the case that the dental nanobots that ought to have gone to me are being illicitly withheld.

As we shall see in Chapter 6, the poor are not stupid. They are aware that they inhabit a world in which others have more than enough. They watch TV shows that encourage them to vicariously enjoy the lifestyles of the wealthy. We have no such direct contact with those enjoying the hedonic benefits brought by twenty-third-century well-being technologies.

How to make comparisons that best reveal the effects of technological progress

Ridley's comparison ignores hedonic normalization. But Gibbon's comparison seems problematic too. His claim about the great happiness of Antonine Rome is simply false if presented as a claim about the average well-being of all the people in the Roman Empire. Gibbon had the selective vision on well-being that would have been typical of his social class. He was interested in the Roman patricians who would have been the equivalents of the ruling-class Englishmen who populated his circle of friends and acquaintances. Gibbon's comparison is difficult to square with the fact that approximately a third of the population of the Roman Empire were slaves. Some of these were comparatively well-looked-after domestic slaves. But many were agricultural slaves who were basically worked to death. There were many injustices in the England of the mid to late 1700s, but none whose magnitude matched this one. The people of the London and Manchester of Gibbon's day did not depend on vast slave estates for their food. England was undergoing an agricultural revolution that was dramatically increasing farm productivity, bringing to an end historical cycles of famine.

For Gibbon's statement to stand any chance of being true it must be interpreted as a claim about the groups with the highest social standing and therefore—we can presume—the best access to the well-being technologies of the day. We can understand him as reporting that the members of the Roman ruling classes had a higher average well-being than did the members of the ruling classes of Gibbon's day. Gibbon might have offered little more than class prejudice to justify his focus. The stories of the impoverished and powerless of ancient Rome are not objectively less interesting than are the stories of their masters and leaders. But we can offer something more theoretically principled to justify focusing on members of the upper social strata in both Antonine Rome and Gibbon's England.

Our interest is in the consequences of technological progress for subjective well-being. It is by focusing on the groups with the best access to well-being technologies that the effects on well-being specific to technological progress become most apparent. We would be comparing citizens of ancient Rome whose houses were internally heated, who had access to the best fruits of Roman agriculture, and whose health was monitored by doctors with the best available understanding of medical science, with the citizens of Gibbon's England who were the best housed, the best nourished, and had access to the best medical care. We would understand Gibbon as asserting that the Romans of 100 CE with the best access to the well-being technologies of their day were happier than the Europeans of 1776 with the best access to well-being technologies of their day.

This way of making comparisons neglects a great many issues relevant to moral thinking about technology. For example, we should distinguish questions about the value of technological progress from questions about the distribution of currently existing technologies. This book is concerned primarily to reject exaggerated claims about the value of technological progress. These claims are distinct from claims about just distributions of existing technologies. Antonine Rome was a time of great distributive injustice. These injustices extended to access to well-being technologies. This prompts the question of what we should do about such injustices. The radical optimists find the solution in accelerating technological progress. As will become clear in Chapter 6, my scepticism about the value of technological progress suggests that it is more important to address injustice than to further accelerate technological progress.

Members of the highest social strata of Antonine Rome can be presumed to be receiving the maximum benefit from the well-being technologies of their day. Members of the highest social strata of Gibbon's England can be presumed to be receiving the maximum benefit from the well-being technologies of their day. A comparison should reveal a difference in the power of the two eras' well-being technologies. We compare like with like. It would be misleading to seek to compare the well-being of a Roman agricultural slave with that of a flourishing inhabitant of early twenty-first-century Manhattan. And it would be misleading to seek to compare the well-being of a flourishing Roman patrician with that of an unemployed, homeless early twenty-first-century Manhattanite. We must compare flourishing Manhattanites with flourishing Roman patricians. If Gibbon is right, then we will find that the patricians experience higher well-being than the nobility of eighteenth-century England. We can assume that an early twenty-first-century Gibbon might have made an analogous claim about the patricians of late first-century Rome and wealthy early twenty-first-century Manhattanites.

Complete or incomplete hedonic normalization

We have noted that hedonic adaptation is biologically adaptive. Humans would achieve few biologically important goals if our first significant success produced a level of bliss that did not fade with time. From natural selection's perspective, a somewhat intense burst of bliss that rapidly fades is better. The intense emotion does not hang around long enough to impede the hard work required to repeat it.

As we saw in Chapter 3, there is a range of views about the completeness of hedonic adaptation. The metaphor of the hedonic treadmill suggests that one's level of subjective well-being is invariant in the face of fortune and misfortune. I proposed that the evidence supports a view according to which hedonic adaptation is often incomplete. There is scope for an analogous dispute about the completeness of hedonic normalization. Hedonic normalization occurs through the internalization of facts about an individual's environment so as to make hedonic rewards and punishments biologically adaptive. If the system for conferring hedonic rewards pays out for trivial achievements then there is a risk that the individual will not achieve his or her biological potential. If the system demands too high a degree of achievement for a hedonic reward then an individual will not be sufficiently rewarded for achievements that have made a positive difference to his or her biological fitness.

Suppose that human psychology evolved for the conditions of the Pleistocene. The psychological and emotional mechanisms responsible for producing different levels of subjective well-being ought to be capable of responding adaptively to typical Pleistocene environments. This is an assumption of evolutionary psychology. We can define complete hedonic normalization in the following way.

> *Complete hedonic normalization*: a fit between an individual's environment and hedonic rewards that is as close as that typically realized in the environment of evolutionary adaptedness (see Fig. 5).

When complete hedonic normalization occurs, the hedonic reward system works as well in a new environment as it did in the environments for which it evolved. When hedonic normalization is incomplete the match between environment and hedonic rewards may fail to be as close as that realized in the Pleistocene (see Fig. 6). A gap can open up. Significant improvements to human environments, such as those produced by technological progress, may result in greater hedonic rewards than those experienced in the Pleistocene.

Hedonic normalization could always be complete. Or it may be complete in some environments but not in others. If hedonic normalization is always complete then it should make no

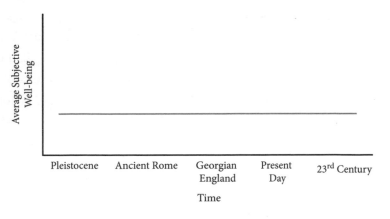

FIG. 5 Complete hedonic normalization.

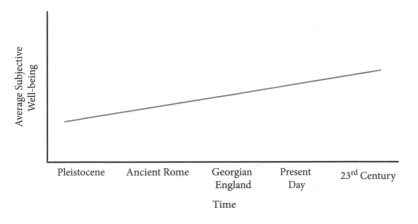

FIG. 6 Incomplete hedonic normalization.

difference to your expected subjective well-being when you are born. Those born into first-century Rome and those born into twenty-first-century Rome will have lives that are different in many ways. But these differences should not affect their expected level of subjective well-being. People who come to maturity in first-century Rome would form goals that match the conditions of that

87

time exactly as well as the goals formed by Pleistocene humans matched the conditions of their existence. Their evaluation of their achievements will be informed by the well-being technologies of that time. People who come to maturity in twenty-first-century Rome form goals and experience hedonic rewards that are perfectly responsive to the well-being technologies of the later age. People born into twenty-third-century Rome will have expectations and hedonic rewards perfectly responsive to the technologies of that age.

Why hedonic normalization is probably incomplete

The conclusions of this book do not depend on the completeness of hedonic normalization. They depend on a propensity for hedonic normalization to modify the contribution that hardships and successes make to subjective well-being. I suspect that hedonic normalization is typically incomplete. An implication of imperfect hedonic normalization is that it is generally better to be born at a time with more rather than less advanced well-being technologies.

Advances in a variety of well-being technologies have done much to reduce the extent and frequency of suffering. Consider advances in medical technology. We now have effective treatments or cures for diseases that caused great suffering to our ancestors.

The phenomenon of hedonic normalization means that we must do more than just tally different eras' unpleasant experiences, thwarted life plans, pleasant experiences, and satisfied life plans to correctly compare the subjective well-being of people living in these times. The mechanism of hedonic normalization leads tooth decay and its consequent pain, when experienced by people in first-century Rome, to have a more modest impact on subjective well-being than the same pain experienced by someone in twenty-first-century Rome. We can imagine that early twenty-first-century root canal procedures would horrify twenty-third-century Romans who will accept as normal dental nanobots that painlessly and perfectly repair any tooth decay. People of different historical eras with different well-being technologies have different dispositions to normalize suffering. This difference in normalization makes a difference to suffering's effect on well-being.

Incomplete hedonic normalization makes it better to be born into a society that has more advanced well-being technologies. Hedonic normalization reduces the suffering permitted by primitive well-being technologies. But it does not reduce this suffering to zero. Smallpox was a normal fact of life in eighteenth-century Europe. It significantly reduced well-being even if it was accepted as a normal part of life. These observations suggest a historical pattern of incomplete hedonic normalization.

There is a further reason it is better to be alive later in a society undergoing technological progress that depends on that progress accelerating. Suppose that technological progress proceeds in a non-accelerating fashion. Suppose further that you have an eighty-year lifespan that you can choose to lead at any time in history. Once you have placed yourself in time you will forget everything about having made the choice. You will be unaware of the fact that you could have chosen to be born into a society with more advanced well-being technologies. You know that the process of hedonic normalization will lead you to accept as normal the circumstances into which you come to maturity. If technological progress does not accelerate then it should deliver as many advances in well-being technologies if you choose to live your life in the first century as if you choose the twenty-first. Suppose now that technological progress is exponential. The eighty-year stretch from 2000 CE to 2080 CE should contain many more technological improvements than the eighty-year stretch from o CE to 80 CE. People who live later benefit from more significant accumulations of hedonic residues. We therefore have a reason to prefer a later life to an earlier life.

The new paradox of technological progress

Hedonic normalization ensures that our hedonic set points are significantly influenced by the circumstances we experience as we come to maturity. Herein lies a difference between improvements of subjective well-being that may result from technological progress and the other goods resulting from technological progress.

Hedonic normalization prevents the straightforward accumulation of the improvements of subjective well-being from technological progress. This means that technological progress has effects on subjective well-being that differ from its effects on other goods. It's a feature of technological progress in general that its benefits accumulate intergenerationally. One generation of farmers discovers a technique that improves the yields of their fields. This technique is passed on to the next generation of farmers who make improvements. A third generation of farmers makes further improvements. The fourth generation inherits not only the improved farming technique, but also the improved yields. The improvement of these yields is not discounted with time. Unless there is some ecological degradation or disaster, a novel farming technique should be no better for the generation that discovers it than for the generation that inherits it. The phenomenon of hedonic normalization means that this pattern does not obtain in respect of well-being. Well-being technologies improve. An individual tends to become happier through the accumulation of hedonic residues impervious to adaptation. But the environmental relativity of subjective well-being prevents these hedonic residues from accumulating intergenerationally. There is no mechanism that transfers a mother's accumulated subjective well-being to her son.

Hedonic benefits from technological progress accrue to individuals at a pace that is significantly faster than the pace at which they accumulate in societies to which these individuals belong. The extent of this gap depends on the efficiency of hedonic normalization. If hedonic normalization is complete then significant benefits should accrue to individuals without having any long-term effects on average well-being. Suppose that hedonic normalization is typically incomplete. Technological progress does tend to produce long-term improvements of subjective well-being. But the gradient of that improvement is not as steep as the gradient of improvement experienced by individuals in their own lifetimes.

This is the error of radical optimism. The radical optimists expect that individuals' hedonic benefits from technological progress can be extrapolated into the future. Indeed, the thesis of radical optimism should lead us to expect an acceleration of the improvement of subjective well-being from technological progress. But this does not occur. Hedonic normalization prevents us from passing on to our children the hedonic residues that accumulate over the course of our lives.

The new paradox of progress is a consequence of a replacement effect. There is turnover in a population. Individuals die and individuals are born. Death eliminates accumulated adaptation-proof hedonic residues. Births introduce to the population individuals whose hedonic set points respond to the well-being technologies that they experience as they mature.

This phenomenon is not restricted to technology. The introduction of democracy to a formerly despotic society should bring greater hedonic returns to the first generation to experience it than to subsequent generations who are hedonically normalized to this social arrangement. The discrepancy is more pronounced in technological progress than it is in other forms of progress quite simply because there is so much of it. For example, I'm not aware of anyone who claims that moral progress is exponential.

One way to defeat the replacement effect would be for humans to become both immortal and infertile. Immortal beings would continue to accumulate adaptation-resistant hedonic residues. Their infertility would prevent the addition of beings whose hedonic set points are normalized in response to the well-being technologies in existence when they join the population. But failing the advent of an infertile immortality we are stuck with hedonic normalization. Deaths and births will continue to press the reset button on well-being.

Figure 7 depicts two different patterns of increase in subjective well-being. Line A represents the increase in well-being experienced by individuals as a result of technological progress. It is formed by the accumulation of hedonic residues from new technologies. Suppose we nominate a period of history—the years 2000 to 2010. We could make a list of all of the new well-being technologies that emerged during that period. Our list would include improved cars (the Toyota Prius), improved mobile phones (the iPhone), improved medicines (face transplant operations and a vaccine for the human papilloma virus), improved jetliners (the Airbus A380), and so on. These will have left adaptation-resistant hedonic residues of varying magnitude in the lives of all of those who experienced them. Were the members of the

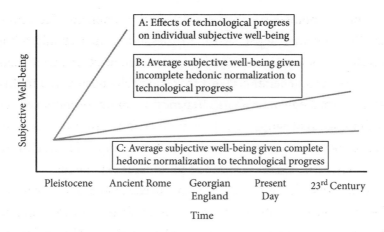

FIG. 7 Comparing improvement of individual well-being with improvement of average well-being.

population to be fixed—all members of the population were immortal and no individuals were to join the population—then Line A would straightforwardly represent the population's pattern of improvement of subjective well-being that results from improved well-being technologies.

However, during the years 2000 to 2010 some people died and some people entered the population. When an individual dies her accumulated hedonic gains from technological progress are removed from the population. The phenomenon of hedonic normalization prevents these gains from passing to a new member of the population. People who come to maturity in a population that has received the benefits of technological progress tend to treat as normal the well-being technologies of their time. These technologies partly constitute their hedonic set points. They begin to experience hedonic improvements from a baseline shaped by the well-being technologies with which they are familiar. Line B represents average subjective well-being from technological progress in a population subject to replacement on the assumption of incomplete hedonic normalization. Line C assumes complete hedonic normalization.

The key point is that A's gradient is steeper than those of B or C. This is a direct consequence of replacement and hedonic normalization. At any given time, subjective well-being is increasing at a rate that significantly exceeds its overall trend.

The combination of the effects of technological progress on individual and average well-being produces something akin to the famous paradoxical stairs of the Dutch artist M. C. Escher. In Escher's picture, a staircase always ascends only to continually return to the point that commences its ascent. Thus we have continued ascent without any gain in altitude. The staircase as drawn by Escher corresponds to the scenario depicted in Figure 8. Individuals experience ongoing steep increases in well-being but continued replacement and hedonic normalization prevents these from having long-term effects.

I suggested that hedonic normalization is likely to be incomplete. Even an artist of Escher's genius would find the corresponding staircase difficult to draw. Those climbing the staircase would actually ascend. But the extent of this ascent would be less than they experience as they climb.

I have presented the new paradox of progress in terms of the effects of more powerful well-being technologies on individual and on long-term population well-being. There is also a temporal dimension. The longer the time period, the weaker will be the effects of a given technological advance on population well-being.

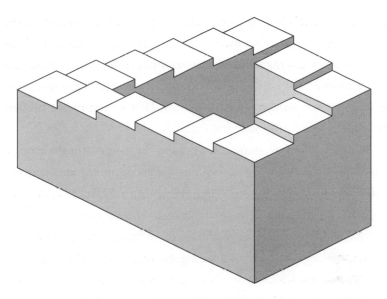

FIG. 8 Impossible staircase.

Consider the six-month period beginning with the introduction of a new well-being technology. The replacement effect and hedonic normalization will have a comparatively small influence over this period. Relatively few people will die and relatively few people will join the population. The effect of the technology on population well-being should not dip too far below the effects on individual well-being. The further in time we are from the introduction of a more powerful well-being technology the greater the percentage of people who view it as normal. Many of the hedonic benefits of a new well-being technology expire. The greatest hedonic benefits brought by the advent of the automobile have, by now, expired. There has been near complete hedonic normalization in respect of the car. We might, in contrast, be experiencing the height of our hedonic returns from the smartphone. Many people have clear memories of the limitations imposed by the emphatically non-mobile rotary dial telephone. We should see a decline in these hedonic returns as the members of a population become increasingly hedonically normalized to the smartphone.

The fact that the process of normalization tends to reduce the hedonic returns of new well-being technologies does not mean that we should seek to do without the technology. With hedonic normalization comes dependence. Perhaps there will be a time when we all view the smartphone as a normal background condition for our lives. Comparisons between life with a smartphone and life without it will no longer seem psychologically and emotionally salient. But we would not want to do without the smartphone— unless, that is, we have a replacement well-being technology that performs all of its tasks at least as well. We might think of our dependence on well-being technologies as somewhat akin to a collective addiction. When we are initially exposed to the new technology it boosts our well-being quite significantly. These returns decline. But we find that we cannot do without it.

These points about declining hedonic returns have some relevance for the discussion about climate change. Many of the harms of climate change are delayed. The generation that is the first to reap the benefits of a new carbon-emitting technology tends not to experience the harms inflicted by it. Future generations are expected to foot the bill for the excesses of our and previous generations. The phenomenon of hedonic normalization places this asymmetry into starker relief. It exposes a selfishness inherent to technological progress. The benefits of a new technology go principally to those alive when the technology is introduced. Later generations are hedonically normalized and therefore receive these benefits only in a significantly discounted form. However, they may have an unmodified experience of any progress trap triggered by a powerful new well-being technology.

Concluding comments

The new paradox of progress applies to enhancements of subject-ive well-being achieved by progress in well-being technologies. It supposes that technological progress can bring about genuine improvements of individual subjective well-being. The paradox lies in the failure of these improvements to carry over at the same level to the population to which individuals belong. The phenom-enon of hedonic normalization significantly limits the extent to which one generation's enhancements of subjective well-being due to technological progress can be passed on to the next generation. The paradox is not limited to technological improvement—it is a feature of any influence that systematically improves the subjective well-being of individuals. We should see it in increases in wealth, for example. There is an often-made observation that the children of people who acquire great wealth are generally better at spending it than further adding to it. This makes the children seem feckless. But we should not overlook the fact that those who make the money tend to experience the greatest hedonic returns from it. The children of the rich treat the conditions of wealth as normal. The magnitude of the effects of technological improvement on well-being accentuates the paradox. The pace of improvement of well-being technologies is accelerating and this is why technological progress manifests the paradox at its most acute.

In the chapters that follow I investigate some practical implica-tions of the new paradox of progress.

5

WE NEED TECHNOLOGICAL
PROGRESS EXPERIMENTS

A focus on subjective well-being enables us to see that the radical optimists overstate the benefits of technological progress. New well-being technologies bring considerable hedonic benefits to individuals. Radical optimists mistakenly extrapolate from these to considerable long-term benefits for the societies to which those individuals belong. The phenomena of replacement and hedonic normalization create a large gap between benefits to individuals and benefits to societies that undergo technological progress.

This chapter switches focus from the benefits of technological progress to its dangers. The historian Ronald Wright offers a view of the risks of progress that serves as a perfect foil for radical optimism. According to Wright, technological progress brings a series of progress traps whose catastrophic consequences increase in magnitude together with increases in technological prowess. We seem presently to find ourselves in the middle of a slow-motion tumble into the progress trap of the climate crisis. This consequence of past and present industrialization has the potential to wreck our global civilization. Its worst-case scenario is human extinction.

This chapter is not a contribution to the voluminous literature on the climate crisis. My focus is on more general questions about how best to achieve the benefits of technological progress while minimizing its dangers. How do we break the cycle of increasingly disastrous progress traps that seem to be features of significant technological advances? Suppose that both our species and our civilization survive the climate crisis. How are we to avoid the progress trap after that? And the one after that?

The benefits of technological progress are overstated. But they are nevertheless genuine. This chapter proposes a way to pursue the benefits of progress while reducing its dangers. We accept a slower pace of technological progress and, in exchange, we reduce the probability of civilization-ending catastrophes.

I make two proposals. The first involves a change in mindset. We must rethink our attitude towards technological progress. Thus far technological progress has always been driven by a competitive ideal. The members of different societies view themselves as competing with other societies to find the most powerful technologies. We should grant a greater role to a cooperative ideal of technological progress. The members of different societies should come to view themselves as collaborating to discover the technologies that best and most safely enhance human well-being. Cooperation at the societal level can coexist with competitive relations between individuals and corporations. Societies will seek to regulate this competition in ways that are consistent with a broader cooperative approach. I will need to distinguish the cooperative ideal of technological progress from wishful thinking. The cooperative ideal is worth striving for. We can make meaningful progress towards realizing it.

The chapter's second proposal is practical. Suppose a group of societies is convinced that the dangers of technological progress warrant cooperation. How are they to realize the cooperative ideal? I propose an experimental approach to technological progress. We should conduct a wide range of progress experiments each of which corresponds to a different potential avenue of technological progress. These experiments are broad. They seek to test not just particular versions of a given technology (e.g. a specific design of nuclear power plant), but rather broad avenues of technological progress (e.g. that comprising technologies that generate power by nuclear fission). Progress experiments are especially important for categories of technology whose longer-term consequences are both uncertain and great in magnitude. They require us to encourage and nurture variation in the paths of technological development. Societies that participate in progress experiments will pursue self-consciously divergent paths of technological development. They will appeal to their own beliefs and values for guidance on which potential avenues of progress to embrace and which to spurn. But they will do so in a way that is informed by the different paths pursued by other societies. They will stand ready to adopt technologies depending on the outcome of the progress experiment. We anticipate some future time when we can decide whether a particular avenue of technological progress can be safely globalized.

Experiments in technological progress offer advance warnings of catastrophes threatened by incomplete understanding of a new, powerful technology. The experimental approach will not work for all varieties of technological progress. It works best for forms of progress that involve the construction of large-scale infrastructure requiring high levels of popular support. It works best for forms of technological progress about which members of different societies tend to endorse different views. I propose that forms of progress represented by nuclear power and genetically modified crops satisfy both these conditions. They represent ideal opportunities for technological progress experiments.

Technological progress traps

This chapter focuses on a danger of technological progress that is well juxtaposed with radical optimism. This is because the aspect of technological progress that makes it dangerous is precisely the one that the radical optimists expect to produce benefits of increasing magnitude.

In his 2005 book, *A Short History of Progress*, Ronald Wright presents technological progress as generating a series of traps, a disturbing feature of which is that we become aware of them only when it is too late to avoid them. This would not be so bad if the chaos and collapse caused by traps did not tend to increase with the power of the technologies that lead to them. But they do. Wright summarizes his view of technological progress by means of a piece of graffiti that he recalls seeing: 'Each time history repeats itself, the price goes up.'[57]

History contains many examples of humans discovering new means to exploit their environments. The results of these forms of exploitation are so attractive that others rush to adopt them. We find ourselves unable to stop until we reach a point at which the exploited resource has collapsed. A resource that seems to its earliest exploiters to be infinite turns out to be finite. Hunting is an early progress trap. Wright says:

Palaeolithic hunters who learned how to kill two mammoths instead of one had made progress. Those who learn how to kill 200—by driving a whole herd over a cliff—had made too much. They lived high for a while, then starved.[58]

Wright adds that: 'The *perfection* of hunting spelled the *end* of hunting as a way of life.'[59] Farming took hunting's place. However, farming brings its own progress traps. Advances in Roman agriculture fed an empire but also produced the deforestation and desertification that facilitated Rome's collapse. Jared Diamond's 2005 book *Collapse: How Societies Choose to Fail or Succeed* documents the many failures of farming that seem to be direct consequences of its success.[60] His examples of the collapse of civilizations include the Easter Islanders, the Greenland Norse, and the Mayans, all of whose preferred forms of progress were improperly informed by the limits of their natural environments.

We now have solutions to past technological progress traps. Were the emperors of third-century Rome to have had access to and to have taken seriously the advice of early twenty-first-century agronomists they could have avoided widespread desertification and deforestation. However, a recurrent feature of the history of technological progress is our failure to work out how to avoid progress traps in time to actually avoid them. Solutions to the problems created by Roman farming techniques arrived soon enough to inform historians' commentaries on Rome's failure but not in time to save the actual Romans. Our persistent failure to avoid progress traps offers inductive support for our inability to escape the progress trap of climate change.

A key difference between past progress traps and the progress traps resulting from industrialization lies in their reach. When Rome collapsed, other civilizations were on hand to fill the vacuum and learn from its mistakes. Wright says: 'The fall of Rome affected tens of millions. If ours were to fail, it would, of course, bring catastrophe on billions.'[61] Wright warns of an impending 'age of chaos and collapse that will dwarf all the dark ages in our past'.[62] Frightening stuff.

The radical optimists are not blind to the perils of progress. David Deutsch proposes that we accept that 'Problems are inevitable, because our knowledge will always be infinitely far from being complete.' He allows that 'Some problems are hard,' but reassures us that 'it is a mistake to confuse hard problems with problems unlikely to be solved. Problems are soluble, and each particular evil is a problem that can be solved.'[63] According to Deutsch, the inevitable missteps of technological progress are no problem so long as technological improvement continues. Tomorrow's inventions will clean up today's messes. Problems predictably caused by tomorrow's inventions will be fixed by the inventions of the day after tomorrow. And so on. The real danger is not from too much technological progress. Rather it's from a failure of nerve regarding its potential benefits. Successfully calling a halt to technological progress would effectively prevent the problems that tomorrow's technologies will predictably cause, but it would leave us without an answer to the problems of the Industrial Revolution which we are now too late to prevent. Deutsch urges us to become an optimistic civilization. An optimistic civilization acknowledges no insuperable problems, no barriers to progress that cannot be overcome.

Suppose that Deutsch is right. Every problem created by technology is soluble by humans. There remains the issue of whether we can find solutions to these inevitable problems in time to save ourselves. When the famous escape artist Harry Houdini was being lowered, feet manacled, into a water-filled tank for his celebrated Water Torture Cell Trick he should not have been overly cheered by the fact that a means of escape could be found—in principle. He needed confidence that he would find it before actually drowning. There are effective technological responses to past progress traps. Nothing prevented the ancient Romans from discovering some of the ecological principles that would have enabled a very productive, but environmentally sustainable agriculture. Deforestation and desertification could have been prevented. But the necessary knowledge came too late to save the Romans. It seems apparent that as technologies become more powerful the progress traps that they generate become more complex. They seem a bit like sudoku puzzles whose complexity increases with increases in the cognitive abilities of their solvers. Can we be so sure that the required advances in scientific and technological understanding will arrive in time to prevent many of the devastating potential effects of climate change? It will be scant comfort to us if the solution is apparent only to extraterrestrials surveying the wreckage of our civilization.

Imagine that Deutsch has placed the winning bet in respect of climate change. This could be the case if we are about to invent miniature robots that, once pumped into the atmosphere, restore greenhouse gases to levels that existed prior to the Industrial Revolution. We have only to keep calm and carry on. But if Deutsch is right that problems are inevitable, then the problem of excess atmospheric carbon dioxide will not be the last civilization-spanning problem we will encounter. Future technological advances may bring some problems that we recognize as essentially the same as those resulting from earlier advances. If a technological advance in the mid-twenty-first century threatens widespread desertification then an understanding of the errors of Roman agriculture may offer some pointers. But we should expect some problems to be genuinely novel. Can we be so sure that Deutsch has bet correctly in the next technological gamble, supposing its potential for harm is, like the climate crisis, truly global? And what about the progress trap with truly global reach after that one? In each case, solutions to problems of progress that are discoverable in principle must actually be discovered before they destroy our civilization.

Note that the problem remains even if we suppose that progress in scientific understanding and technological capabilities gives us a systematically enhanced ability to avoid technological progress traps. So long as novel technologies bring novel dangers and the dangers of technological progress are global, it is inevitable that, at some point in the future, we will place a losing bet. We are in the position of a gambler who continues to bet her entire stake against an infinitely wealthy house. Deutsch's optimism suggests that all of her bets will be well chosen. Even so, it is statistically inevitable that eventually she will lose everything.

We have potential harms from technological progress that we can match with the benefits promised by the radical optimists. The radical optimists acknowledge no limit on the potential increase in benefits from technological progress. The good news is that there could be an upper limit to the harms caused to humans by our technological progress. The bad news is that the upper limit is human extinction. In Chapter 4 I argued that hedonic normalization reduces the magnitude of benefits from technological progress. There is no possibility for hedonic normalization to reduce the harms of extinction. Extinction cannot be a normal condition of any species' existence.

Two ideals of technological progress

This chapter offers a way to get advance warning of technological progress traps. First, it requires a change in mindset. Thus far our species has taken a mainly competitive approach to technological progress. This is especially obvious in the case of weapons technologies. Societies whose soldiers were the first to be equipped with a powerful new weapon exhibited a tendency to invade, enslave, or colonize their neighbours. This pattern creates a powerful incentive to get the technologies first. But there is competition to progress faster in well-being technologies too. Societies whose industrialists are the first to construct powerful new well-being technologies receive significant economic benefits.

The cooperative ideal suggests that societies work together to discover the technologies that best and most safety enhance human well-being. It assumes a collective interest in determining the avenues of technological progress that bring long-term enhancements of well-being without creating progress traps. Is this practical? An ideal of cooperation in technological progress might seem like the ideal of group hugs as a solution to conflict in the Middle East—a nice but dangerously naïve idea.

How should we decide whether the cooperative ideal of technological progress is wishful thinking or warrants serious advocacy? To be worth advocating the ideal should be efficacious. By 'efficacious' I mean that those who defend the ideal should have some realistic expectation that they can bring human behaviour into greater conformity with it. They accept that many people will deviate from the ideal. But the ideal does not have to be maximally efficacious to be worth advocating. Many efficacious ideals nudge our behaviour in the right direction without any realistic prospect of perfect universal conformity.

Consider another ideal frequently dismissed as impractical. Those who advocate the ideals of honesty and transparency in politics needn't be so naïve as to believe that politicians could or should always be maximally candid. They can acknowledge that sometimes there's no alternative to deception. Honesty matters, but it's not the only thing that matters. Sometimes it should be trumped by considerations of political expediency. These concessions do not prevent the ideal of candour from serving as a standard against which to judge the behaviour of politicians in democratic societies. Those who advance the ideal of honesty and transparency hope that its advocacy will bring politicians to conform more closely with it. This is how we can think of the cooperative ideal of technological progress. Advocacy of the ideal can be efficacious. It can influence our approach to technological progress even if the competitive ideal remains more influential.

Some ideals require universal compliance to be successfully implemented. Universal love would be a solution to the world's international conflicts. But it will have few of the desired effects if practised only by a small number of people. It is hard to imagine how a handful of peace campaigners who arrive at the Korean demilitarized zone determined to implement universal love could do anything to settle the dispute between North and South. The implementation of other ideals can start small. They are not as effective when partially practised as they would be if practised universally, but partial compliance does have some of the desired effects. An ideal's efficacy can increase. Practical moves towards universalization can follow from observations of this positive effect. Cooperation in technological progress can begin with baby steps. The cooperative ideal can produce good effects even if it is, at the outset, pursued only in respect of a narrow range of technologies, and even then imperfectly. Evidence of this effectiveness could lead the ideal to become more influential and more widely adopted.

The competitive ideal has produced rapid technological progress. Competition between societies is most intense during times of war. The race for decisive weapons in World War Two significantly accelerated the development of the jet, the computer, nuclear fission, and the rocket. These advances, in turn, accelerated progress in a variety of well-being technologies. The cooperative ideal described in this chapter slows the pace of technological progress. It distributes the world's scientific minds and research dollars across a wide range of possible avenues of progress rather than concentrating them on the few technologies that seem likely to be the most powerful.

The cooperative ideal of technological progress does not require cooperation at all levels. Much of the process of discovering and bringing new technologies to market may be competitive. Individuals can continue to compete with each other to be the first to discover a technology and therefore to reap the reputational and financial benefits. Apple and Google can continue to compete for the profits that accrue to the company with the best mobile mapping software. Cooperation is required at the level of the state, the entity that lays down the rules for competition between individuals and corporations. It applies to the relations between different societies that may be exploring different avenues of technological progress.

The fear of falling behind

One obstacle to cooperation in technological progress is a pervasive fear of falling behind technologically. According to a popular narrative, when, under the Ming Dynasty, China chose to fall behind technologically it was, in effect, asking to be colonized a few centuries later by technologically superior Europeans. The cooperative ideal requires that we cease to view technological progress as involving competitions from which winners emerge triumphant and dominant, and losers emerge defeated and dominated.

The dangers of falling behind are especially salient in respect of weapons technologies and other technologies whose purpose is to enable one culture to dominate another. When Maori warriors confronted British troops they swiftly learned which of the *mere*, a leaf-shaped short club typically fashioned out of greenstone, or the Enfield rifle was the superior weapons technology. The backwardness of their military technologies led rapidly to their colonization. Maori who sought to resist their colonizers had no option but to trade in their clubs for guns.

It would be nice to think that the central narrative of technological progress today involves new means to make people happier rather than more effective means to kill and dominate them. Steven Pinker's important discussion of the decline of violence in his *Better Angels of our Nature* offers some support for this view.[64] Pinker charts a general decline of violence in human societies and a reduction in the frequency and hideousness of war. If Pinker is right, we should be seeing a corresponding reduction of interest in technologies of destruction and domination. To the extent that the future focus of technological progress concerns how best to improve well-being, the cooperative ideal should be increasingly important.

The danger of falling behind is less frightening in respect of well-being technologies than it is in respect of weapons technologies. The members of a society with better plumbing and better cancer medicines may he happier. But these superior well-being technologies do not directly enable the domination of a less technologically advanced neighbour.

This book has suggested that technological progress is capable of boosting well-being. But the new paradox of progress suggests that the enhancements of well-being from technological progress are less than we tend to think. We should recognize that technological progress is one among many influences on the well-being of a society's citizens. It deserves no primacy among these influences. Furthermore, technological progress comes with dangers absent from other positive influences on well-being.

How is progress dangerous?

Before we proceed we need a better understanding of the dangers posed by progress in well-being technologies. Many of Ronald Wright's examples of progress traps involve well-being technologies. Well-being technologies including cars and jetliners are likely to have been significant contributors to anthropogenic climate change. Roman agriculture was, as we noted earlier in this chapter, a collection of well-being technologies that produced widespread deforestation and desertification.

It is important to distinguish the threats posed by well-being technologies from threats posed by progress in other types of technology. Very powerful weapons technologies threaten civilization-ending catastrophes by working precisely as they are designed to work. Thermonuclear bombs are designed to produce mass destruction. It's not surprising that the detonation of just a few of them could bring our current civilization to a direct and decisive end. The use of many devices designed to create destruction should predictably create a great deal of destruction. The threat from progress in well-being technologies is different. Deforestation and desertification were not designed effects of Roman agricultural technologies. Contributions to atmospheric carbon do not appear on the list of results sought by the designers of cars or coal-fired power plants. They are unanticipated effects that become apparent only once the manifest power of the technology has caused it to spread throughout an adopting civilization. Catastrophes result when an accumulation of these unanticipated effects exceeds a certain threshold. The danger from well-being technologies comes from a premature universal adoption insufficiently informed about longer-term effects.

We should distinguish progress traps from accidents. The 1986 Chernobyl nuclear accident was tragic. But, in isolation, it did not threaten the end of modern industrial civilization in the way that an exchange of nuclear weapons would. The mass adoption of nuclear power could threaten the end of civilization through the insidious accumulation of some effect that we do not notice until our global industrial civilization finds itself irrevocably committed to nuclear power. Perhaps there are unanticipated global dangers in the storage of large quantities of nuclear waste. This is how it would qualify as a potential progress trap.

We strive to prevent civilization-ending uses of nuclear weapons by keeping them away from individuals and groups disposed to use them—or at least, away from those disposed to make first use of them. Since the potential civilization-ending effects of well-being technologies depend on their universal or wide adoption, we should be able to prevent them by being cautious about their spread. Progress experiments are designed to detect progress traps before we collectively stumble into them. If certain societies adopt new technologies while others do not, we have the variation required for a scientific experiment. The important thing is to resist the global spread of a powerful new well-being technology before the results of the progress experiment are in.

We should self-consciously view ourselves as involved in not just one grand experiment with technological progress as a whole, but instead in a series of smaller experiments to show how best to direct technological progress at the enhancement of human well-being.

Rehabilitating the idea of technology experiments

The word 'experiment' may seem ill chosen. Wright presents our technological civilization as a 'great experiment'. It belongs to a sequence of progress experiments that includes the introduction of powerful hunting technologies and large-scale Roman agriculture. Wright says 'the wrecks of failed experiments lie in deserts and jungles like fallen airliners whose flight recorders can tell us what went wrong'.[65] We now understand how the technological proficiency of Maori moa hunters destroyed the way of life of moa hunting by driving those animals into extinction. We can see how the agricultural innovations of the Romans produced mass deforestation and desertification. In each case we seem to acquire the knowledge to avoid catastrophe only by experiencing the catastrophe. The increases in the power of our technologies—causes for great celebration by radical optimists—are, according to Wright, taking us to 'an age of chaos and collapse that will dwarf all the dark ages in our past'.[66]

Wright's use of the term 'experiment' to describe our civilization's engagement with technology clearly has pejorative overtones. For example, people who weren't aware that the corn that they consume has been genetically modified are told that they were duped into serving as guinea pigs in an experiment on the safety of genetically modified food. But experiments don't have to be this way. We have many examples from medicine of people freely consenting to be experimental subjects in clinical trials that may offer valuable new therapies for diseases. Medical experiments are fraught with dangers, but they can be conducted in an ethically responsible way. Human subjects in well-conducted medical experiments are informed of the risks and reassured that their participation can produce benefits both for themselves and for others with their medical condition.

Experiments in technological progress must be different from those described by Wright. In the experiments he describes, entire cultures or civilizations bet all they have on certain varieties of technological progress. Wright's civilization-spanning experiments differ from well-conducted medical experiments. The latter test a new therapy before it is introduced into the patient population. A medical researcher does not test on the entire patient population a potentially beneficial, but also potentially lethal new medical therapy and call what she has done an experiment. To do so would not be to run an experiment, rather it would be to gamble in an especially morally negligent way. Experiments are supposed to anticipate the universal spread of a new powerful well-being technology.[67]

Jared Diamond on the natural experiments
of traditional societies

How might we design an experiment in technological progress? Jared Diamond claims that we can learn many lessons from traditional societies about how best to raise our children, treat the elderly, nourish ourselves, and resolve conflicts.[68] According to Diamond our understanding of human nature is overly influenced by study of a narrow sample of the ways humans can be or live. Diamond says: 'if we wish to generalize about human nature, we need to broaden greatly our study sample from the usual WEIRD subjects...to the whole range of traditional societies'.[69] Here 'WEIRD' stands for Western, educated, industrialized, rich, and democratic. It does seem to be the case that much of our understanding of human psychology and behaviour comes from investigations of subjects from WEIRD cultures and that some of the conclusions we draw about human nature may not apply to people from traditional cultures. People who come to maturity in modern-day Pittsburgh, Paris, and Perth, are subject to different influences, but the differences are small when compared with the differences between any one of these people and those who come to maturity in a traditional society of the Papua New Guinean highlands.

This bias manifests also in respect of our awareness of potential solutions to our problems. We tend to overlook the discoveries of traditional societies. Diamond says:

> Traditional societies in effect represent thousands of natural experiments in how to construct a human society. They have come up with thousands of solutions to human problems, solutions different from those adopted in our own WEIRD modern societies.[70]

Diamond offers a powerful instrumental reason to preserve much of this diversity in approaches to difficult social problems. Many practices of traditional societies strike us as strange and wrong-headed purely because they differ starkly from what we are used to. In the past, European colonizers rejected traditional practices out of hand. An unbiased assessment of traditional ways of raising children, caring for the elderly, nourishing ourselves, and resolving conflicts may reveal some of them to be superior to WEIRD ways.

Creating and nurturing variation in technological progress

Can we apply Diamond's advice to forms of technological progress? There seems to be a significant obstacle here. A comparison of the well-being of people in WEIRD and traditional societies may help us to answer very general questions about the value of technological progress. But the conclusions of such a study will be of little practical value to the members of WEIRD societies facing tough choices about which potential path of technological progress to follow. Our identification of traditional societies as 'primitive' has the effect of reducing our interest in lessons that they might have learned about technology. We are unlikely to be receptive to lessons about the valuable familial bonding possible in cultures without television if receiving these benefits requires us to give up our TVs. Furthermore, conclusions about the content-ment of people who, in their lifetimes, have accumulated very few hours of TV watching may have little relevance to decisions about how to make WEIRD people happier. The varieties of television-free bliss enjoyed by traditional peoples are unlikely to be available to the citizens of WEIRD societies who are thoroughly hedonically normalized to television.

Experiments require variation. When variation leads to differences in a phenomenon of interest we hope to draw conclusions about the causal relevance of the properties that vary. In the physical sciences this variation is sometimes created in a laboratory. Diamond proposes that we look to traditional societies for naturally occurring variation in child rearing, dispute resolution, and a variety of other practices. An experimental approach to technological progress looks to variation that tends to occur in different WEIRD societies in respect of new well-being technologies. This variation occurs as the citizens of different WEIRD societies respond differently to a new, potentially dangerous well-being technology. If we are to conduct experiments on different avenues of technological progress then we should encourage and nurture much of this variation.

This chapter explores two examples of potential technological progress which have elicited differing responses from the peoples of different WEIRD societies. These involve nuclear power and genetically modified crops.

Many participants in current debates about nuclear power and genetically modified crops seem to suppose that we already know enough to decisively pronounce on the safety of these forms of progress. Alertness to the possibility of technological progress traps should lead them to recognize the value of experiments in technological progress.

The conclusions we draw from experiments in different varieties of technological progress can be compared to those drawn from the testing of new medical therapies on human subjects. Suppose we want to test a new medical therapy. We could perform an experiment in which some participants receive the therapy while others receive a placebo. We seek to draw inferences about the drug's efficacy by observing differences between the two groups. In a progress experiment, some WEIRD nations will embrace a form of technological progress. Others will not. We seek to observe relevant differences between the two categories of society. In a clinical trial it is typically the case that a random process decides who among the patient group receives the compound under examination and who receives the placebo. It would be impractical to give to a body like the United Nations the task of randomly allocating nations to new well-being technologies. Instead we must rely on the values and beliefs of the citizens of technologically advanced nations to decide whether they receive the new technology. New avenues of technological progress bring benefits. They are also associated with dangers. The different values and beliefs of the members of different societies will lead them to weight these benefits and dangers differently. We should resist significant pressure to eradicate the diversity of views about the value of a given category of technological progress. This diversity can act as insurance against the errors in the accepted path of progress.

When we draw our conclusions we must be alert to the relevance of other factors. There are many differences between two groups of WEIRD societies beyond a general tendency to endorse an avenue of technological progress in one and a tendency to oppose it in the other. For example, there may be differences in the safety cultures of different WEIRD nations. Consider a nuclear power progress experiment. Differences in safety cultures rather than anything intrinsic to the technologies of power generation may produce a high rate of accidents in one nation. We would have received a misleading picture of the safety of nuclear power had we been unduly influenced by the experiences of the Soviet Union in the 1980s. We can and should be cautious about the conclusions we draw from any experiment. Those who design experiments on new pharmaceuticals understand that they cannot control for every influence on how an experimental compound works. For example, there are likely to be genetic differences between the members of the group that test the compound and those in the placebo group that will influence responses to the drug. These complications need not invalidate experiments on new medicines, and they need not invalidate technological progress experiments. It is one thing to say that we must take into account facts beyond whether a society endorses nuclear power. It is another to say that we can learn nothing about the safety of nuclear power as a variety of technological progress by studying differences between societies that endorse nuclear power and those that oppose it.

Those who, like Wright, take a negative view of experiments in technological progress tend to present a culture's citizens as unwitting or unwilling guinea pigs. In progress experiments as conceived here, the creation of variation required by an experiment responds to different cultural affinities in respect of a new avenue of technological progress. A democratic process is one way to decide whether a society is in the group that tests a new avenue of technological progress or whether it is in the group that explores alternative approaches. In the sections that follow I will indicate how the democratic process can produce the variation required for progress experiments on nuclear power and genetically modified crops.

Progress experiments offer one way to avoid the disastrous outcomes described by Wright and Diamond. The future is uncertain. When a category of technology becomes truly global we, in effect, place all of our eggs in one basket. Roman agriculture might not have been the catastrophe that it turned out to be had it been restricted to certain regions of the Empire. Watchful imperial administrators might have been in a position to observe the desertification it created in one previously fertile region and warn against its spread.

A nuclear power progress experiment

There is currently an emotionally charged debate about the safety of nuclear power. Nuclear power's opponents point to Chernobyl and Fukushima and warn of the potential for even more disastrous accidents. They say that the Fukushima accident was a result of nature refusing to play along with the ostensibly well-laid plans of the designers of nuclear power plants. The plant's designers had taken into account the possibility of a major earthquake. They had also taken into account the possibility of a tsunami. They appear to have been caught out by the combination of the two that occurred on 11 March 2011. Opponents of nuclear power also worry about long-term solutions for the waste that nuclear power plants generate. The advocates of nuclear power point to the apparent greenness of nuclear power plants. These plants generate large amounts of energy while producing very little carbon. Advocates urge that we not overlook our capacity to learn from past accidents. In the wake of the Fukushima accident, some very intelligent people have been working out ways to protect nuclear power plants from the dual threats of earthquake and tsunami.

The Fukushima accident has prompted some nuclear nations to reconsider their commitment to nuclear power. Possibly the most significant development was an announcement by Chancellor Angela Merkel that Germany would have closed its last nuclear power plant by 2022. Other WEIRD nations, the United States among them, show no signs of renouncing nuclear power.

This has been the cause of intense and vituperative debate. Partisans of both sides of the debate accuse their opponents of sloppy, emotionally driven thinking. The defenders of nuclear power find its opponents to be overly influenced by non-scientific ideologies. Opponents of nuclear power accuse its defenders of being guided by a technophilia that prevents them from giving due weight to the dangers to the lives of ordinary people posed by the plants. They say that defenders of nuclear power fail to consider technologies for generating energy that are safer in the long term but which may do less to boost industrialists' profits.

Both these alternative views assume a claim that the experimental approach to technological progress challenges. They suppose that we know enough to pronounce conclusively on the safety of the avenue of progress represented by nuclear power. This is false. Both sides in this debate should manifest an appropriate epistemic modesty. The experimental approach allows that there is uncertainty about the long-term safety and value of nuclear power. We should conduct an experiment. Opinion polls reveal a variety of views about the value of nuclear power among the peoples of different WEIRD nations. Those interested in conducting experiments in technological progress should encourage this diversity rather than viewing it as an indicator of the foolishness of those on one side or other of the debate. The question of the safety of these varieties of technological progress remains open. People may be prepared to make educated guesses on the basis of the evidence to date. But these guesses can be granted the same status as the guesses of a medical researcher who fully expects her new compound to be an effective treatment. Experienced medical researchers may be quite good at predicting the outcomes of clinical trials. But these predictions certainly do not obviate the need for experiments and the confidence that their successful execution can bring.

The experiment that I envisage does not test designs for individual nuclear power plants. It is broader than this. Its focus is on nuclear power as a form of technological progress. Experiments on individual instances of a new technology are important. But so are experiments that test the value of entire avenues of technological progress.

How might this experiment proceed? Germany is a large, industrialized nation. If Germany renounces nuclear power it will need to find other ways to generate sufficient energy to meet the needs of its citizens and industries. It is now apparent that the strategy of meeting additional demands for power by building more coal-fired power plants is mistaken. Each additional coal-fired power plant exacerbates the problem of climate change. Alternatives to nuclear energy may involve massive wind farms or solar energy plants. These green options have their own disadvantages. Wind farms take up a great deal of land and produce less power if the weather does not cooperate. We can expect creative Germans impressed by their nation's current and projected needs for power to come up with other ways to generate energy that do not involve nuclear fission. Over time, smart people in nations that pursue the nuclear option will seek to make their plants as safe and efficient as possible. Over the same period of time, smart people in the nations that reject the nuclear option will seek to make non-nuclear power plants as safe and efficient as possible.

The experiment must run for some time if we are to avoid the bias of permitting our judgments to be unduly influenced by unrepresentative members of a particular category of technologies. For example, it would not be right for judgments on the safety of nuclear power to be overly influenced by conclusions drawn from the Soviet graphite-moderated reactor of the Chernobyl plant. To do so would be akin to overemphasizing the fatality statistics of the 1961 Lincoln Continental, infamous for its front-opening rear 'suicide doors', in an assessment of the variety of technological progress represented by the automobile. The Continental's doors were prone to open during crashes, hurling an unstrapped-in backseat passenger out of the car. It was easy to replace front-opening doors with safer rear-opening doors. An experiment on the safety of nuclear power as a variety of technological progress should take account of how the designers of nuclear power plants respond to problems that inevitably arise.

How long should an experiment on different approaches to power generation run for? It must last long enough to test a wide variety of ways to make power that either involve nuclear fission or do not. But we should have in mind a future time when it would be appropriate to pronounce on the result of the progress experiment. When asked in the early 1970s whether the French Revolution should be considered a success or a failure, Zhou Enlai is reputed to have answered 'It's too soon to tell'. It would be disappointing if we were still waiting to hear whether the generation of energy by nuclear fission was a success or failure 200 years after the beginning of the progress experiment.

One encouraging feature of the debate is that neither defenders nor opponents of nuclear power are shy about making predictions. The nuclear power progress experiment demands that opponents and defenders of nuclear power commit to specific predictions capable of straightforward falsification by future events. Opponents expect to see more accidents and radioactive leaks. Proponents expect few of these. Proponents might predict a pattern similar to what occurred with attempts to improve the safety of passenger jets. Modern jetliners are complex machines that can go wrong in a variety of ways. Such is their complexity that even the most scrupulously safety-conscious aeronautical engineers cannot anticipate all potential accidents. Generally, they become aware of design faults only by picking through the wreckage of crashed jetliners. We have not seen the last such accident. Jetliners will continue to crash for unpredictable reasons. But the safety record of jetliners is nevertheless encouraging. The frequency of jetliner crashes and other serious accidents has declined significantly since the early days. People who today happily travel long distances in cars should feel confident about covering these same distances in a well-maintained passenger jet. In fifty years time, people might make analogous observations about the safety of nuclear power plants. There will always be unpredicted things that can go wrong with a complex thing like a nuclear power plant. But we may witness a decline in the frequency and severity of these events that justifies an increasing confidence in their safety.

One of the most valuable contributions of the great philosopher of science Karl Popper is his rejection of certainty.[71] According to Popper, the scientific method does not deliver the certain truths that some religions claim to offer. Scientific understanding grows with the accumulation of conjectures about the universe that are yet to be falsified. Our most fundamental scientific understandings have survived many tests. But there is no law of logic that prevents them from falling at the next opportunity for falsification.

We should say something similar about experiments designed to show that a particular technology is safe. There is no such thing as a perfectly safe technology. A technology that has passed all extant tests of safety may be unsafe in ways that we barely understand and therefore cannot currently test for. It could generate a progress trap so fiendishly complex we need major advances in science to begin to comprehend it.

The Popperian approach to scientific claims can lead us to be increasingly confident in them without ever placing us in a position to pronounce them certainly true. Something analogous could happen in respect of tests of the safety of technologies. Suppose we initiate a nuclear power progress experiment. A twenty-year period passes without any serious accident. In the nations that embrace nuclear power the needs of industry are met in a way that creates little pollution. This pattern should be viewed as encouraging. Twenty years is not a particularly long period of time. But we can imagine that some nations on the side of the progress experiment that rejects nuclear power will reconsider. These might be nations in which, at the outset of the progress experiment, there was only a fairly small majority opposed to nuclear power. The citizens of other nations will not be persuaded by a mere twenty-year period of uneventful operation of nuclear power plants. The progress experiment continues, with the ratio of societies on each arm indicating global confidence in the safety of nuclear power. It's probable that memories of Chernobyl and Fukushima will persist longer in the minds of members of some societies. If there are always some societies that reject nuclear power we will continue to have some insurance against the failure of the progress experiment, even if we believe that a negative verdict is increasingly unlikely. There is no time period sufficient to pronounce nuclear power as absolutely safe.

The reverse outcome should see a similar incremental passing of judgment. Suppose that there are more accidents of the type that occurred in Chernobyl or Fukushima in nations that accept nuclear power. Problems recur in spite of the best efforts to make nuclear power plants safe. There might be an increase in defections from the arm of the progress experiment testing nuclear power. If the trend continues only the nations whose citizens are most confident in their scientists' abilities to fix misfiring technologies will stick with nuclear power. But there may come a time when even they are persuaded to renounce energy produced by nuclear fission. When the last nuclear power generating society decommissions its plants, we can say that our global civilization has passed a definite verdict on the safety of that way of generating power. There may be ways to make nuclear power safe that humans were not fortunate enough to discover, but there has nevertheless been a collective decision that nuclear power is too dangerous.

The experiences of WEIRD peoples in the other arm of the trial are always relevant. Suppose that US nuclear power plants become safer. In the meantime, the Germans discover a way of generating power that is both safer and more efficient than the best that nuclear power offers. This would be one way for the progress experiment to conclude against nuclear power. The important thing is that we avoid hasty and biased evaluations of different avenues of technological progress.

The nuclear power progress experiment could yield a more depressing outcome. The energy alternatives explored by the Germans may condemn them to blackouts and brownouts, followed by industrial contraction and mass unemployment. Americans may find that nuclear power plants remain disastrously accident-prone. If this is the outcome of the experiment then we may need to return to the arguments that feature in current debates about nuclear power. It may turn out that the nuclear option is superior in some significant respects and inferior in others. The data from the technological progress trial should nevertheless enable a more informed choice. The democratic process is an ideal mechanism for making such choices.

Why should the winners share with the losers?

It's not surprising that differences in the values of WEIRD societies lead them to evaluate the benefits and risks of nuclear power differently. This is good news for a nuclear power progress experiment. The variation required for an experiment shouldn't be too difficult to generate. What is more difficult to find is a mechanism whereby the knowledge emerging from the progress experiment can be shared.

Suppose that after a sufficiently long period the progress experiment concludes in favour of nuclear power. American nuclear engineers vastly improve the safety of plants. They deal effectively with the problem posed by nuclear waste. The alternatives pursued in Germany prove dirty, dangerous, and incapable of meeting the needs of the citizens of an advanced industrialized nation. Why should Americans let Germans receive the benefits of their shrewd bet on nuclear power? It seems unduly optimistic to suppose that they would simply send over the designs for safe, efficient nuclear power plants as an email attachment.

This chapter advocates a technological progress ideal. One does not straightforwardly falsify an ideal by pointing to a range of cases in which people fail to live up to it. The cooperative ideal should be efficacious. Advocating it should tend to produce greater conformity with its requirements. This is the way we think about cooperation in respect of the climate crisis. There are significant short-term rewards to nations that shirk their obligations in respect of climate change, so it's not surprising that many nations are not doing enough to cut greenhouse gas emissions. But the ideal of cooperating to slow climate change is efficacious. We see halting moves towards cooperation. Doubts remain as to whether this is too little, too late—advocacy of the ideal may not be *sufficiently* efficacious.

Societies that choose nuclear power in the progress experiment should acknowledge that their decision to endorse nuclear power occurs in the context of a collective endeavour to establish whether or not it is safe. They are relevantly similar to individual participants in a trial testing a new medical therapy.

When medical scientists test a new medical therapy on a group of patients, the group receiving the placebo has some entitlement to the tested compound should the clinical trial prove its therapeutic value. They do not receive the tested compound in the clinical trial, but they are right to consider themselves part of the group testing the compound. Data on their response to the placebo forms part of the case for the therapeutic value of the drug. They have no less of an entitlement to the tested compound than do the participants who received it in the trial, should its therapeutic value be demonstrated. Americans should understand that their confidence in nuclear power comes from an experiment in which non-nuclear Germans participated.

It makes sense for Americans to be generous in imparting the lessons they have learned if they understand the non-nuclear alternative as a form of insurance. They should allow that there was a very real chance that nuclear power would turn out to be dangerous. If so, then they hope that the German model would be available to them. Even those very confident of the superiority of nuclear power should be susceptible to this reasoning. People find it rational to purchase insurance for events that are quite improbable. The probability of any given house burning down can be low. Yet an appropriately priced fire insurance policy is typically still a good idea. Americans confident about the safety of the nuclear power plant should be grateful for the existence of the German alternative. They would undertake to supply data to the Germans so as to guarantee access to the German model should nuclear power turn out to be the wrong option.

Offering the results of a progress experiment to the Germans ensures that there can be further experiments in technological progress involving both nations. Technological progress experiments will not be limited to nuclear power. This progress experiment will proceed alongside progress experiments involving a wide variety of new technologies that have the potential to produce significant benefits but also threaten progress traps. Nations that refuse to play fair in one progress experiment will find themselves excluded from others. A nation that refuses to share data on its correct choice in a technological progress experiment would be seen as abrogating its most serious international commitments. Those who recognize the importance of avoiding progress traps will understand that if nuclear power is shown by the experiment to be clean and safe it should be spread. It's better that the knowledge is spread rather than letting other nations blunder their way through a potentially environmentally destructive process of finding out how to make nuclear power safe.

A progress experiment on genetically modified crops

The genetic modification of crops provides a further opportunity for a technological progress experiment. Commentators on this debate note different patterns of response in different parts of the world.[72] In 2012, genetically modified crops were grown in twenty-eight countries.[73] The United States led the way with 69.5 million hectares of genetically modified maize, soybean, cotton, canola, sugar beet, alfalfa, papaya, and squash. It was followed by the 36.6 million hectares of soybean, maize, and cotton planted in Brazil and the 23.9 million hectares of the same crops in Argentina. Canada came in fourth with 11.6 million hectares of canola, maize, soybean, and sugar beet. The Europeans have, in general, been reluctant to follow the new world's lead into a genetically modified future.

According to one interpretation of this debate, the issue of genetically modifying crops is an issue on which communities that reject the technologies of genetic modification and communities that permit them cannot both be right. The bitterly waged argument between opponents and advocates of genetically modified crops seems to suggest such a view. We know enough to recognize that the planting of genetically modified crops is a great idea or foolishly reckless. Rather than seeking to pronounce definitively on this debate now we should acknowledge the conditions for a progress experiment in agricultural genetic modification. There is variation among the citizens of WEIRD nations on the safety of genetic modification. For example, in a survey of perceived dangers of genetically modified food conducted in 1997, 69 per cent of Austrians viewed genetically modified foods as bringing serious risks while only 14 per cent of US respondents did.[74] These numbers are likely to have changed over the years since the poll was conducted. But they suggest that democratic processes should deliver the variation required for a progress experiment.

We should be clear about which claims about genetically modified crops we are testing. There is an ambiguity in the suggestion that genetically modified foods 'bring serious risks'. Some opponents of genetically modified food crops understand these as risks for those who consume genetically modified food. They perceive genetically modified foods as poisonous. They speculate about an increased cancer risk for those who consume genetically modified nachos. A second kind of risk concerns the environmental consequences of the planting of genetically modified crops. This line of opposition points to the fragile complexity of natural ecosystems upon which humans depend. The second kind of opposition finds a variety of expressions in the debate about genetically modified crops. One frequently expressed fear concerns horizontal transfer. This could see the genes deliberately introduced into a crop transferred by bacteria or viruses into other plant species. Opponents fear the accidental creation of genetically enhanced super-weeds.

Of these two fears, the latter is suitable for a progress experiment while the former is not. It's easy to understand where the fear of poisoning comes from. We have an evolved suspicion of unfamiliar foods. Humans in the environment of evolutionary adaptedness who consumed unfamiliar things risked finding themselves poisoned. Some of this evolved suspicion of unfamiliar food seems to have transferred to genetically modified food crops. This point notwithstanding, it is wrong to permit a fear of novelty to condemn an entire category of foodstuffs. Genetically modified food encompasses a wide variety of foods modified in a wide variety of ways. No genetically modified food is absolutely safe. But absolute safety isn't a standard that any food can achieve. The most benign-looking lentil stew could harbour a lethal *E. coli* contamination. We should not demand that a novel food meet the impossible standard of absolute safety, but instead that it achieve the more lenient standard of being as safe as the other foods that we happily consume. Genetically modified foods can achieve that standard.

More importantly, even if there was a substantial risk of poisoning from genetically modified food, that wouldn't count as a progress trap in the sense described by Wright. Progress traps are, by their very natures, stealthy. We are likely to discover the threat of poisoning from genetically modified crops well before the point at which mass poisonings lead to a collapse of our global civilization.

Progress experiments of genetically modified crops are better directed at the environmental consequences of genetic modification. We should acknowledge the complexity of the natural ecosystems into which we introduce genetically modified crops. Defenders of genetically modified crops cite their many advantages over non- genetically modified equivalents. Crops can be modified to be more nutritious, to be more productive, to grow in inhospitable regions of the world, to be resistant to common pests, and so on. These perceived advantages may lead to the global spread of certain genetically modified crops. It is therefore important that we be alert to the possibility of a progress trap. We should grant that we do not fully grasp the complexity of the biosphere. We should therefore allow that there may be some disastrous environmental consequence which becomes apparent only after the widespread planting of genetically modified crops.

Past progress traps offer some inductive support for this concern. Roman imperial administrators are unlikely to have been capable of understanding the ecological consequences of their agricultural practices. The immense complexity of the biosphere leaves plenty of room for dangers from genetically modified crops that are beyond the reach of early twenty-first-century ecology.

One fashionable response to this uncertainty comes in the form of the precautionary principle. According to the Wingspread Statement, one of the most influential versions of the precautionary principle: 'When an activity raises threats of harm to human health or the environment, precautionary measures should be taken even if some cause and effect relationships are not fully established scientifically.'[75] Advocates of the precautionary principle assign the burden of proof to the proponent of a technological novelty. They insist that an examination must not overlook the option of inaction. The precautionary principle sets a high standard of proof that a course of action does not harm human health or the environment. In practical terms, it is likely often to lead to inaction, as no proponent of positive steps is able to discharge the burden of proof that no harm is caused. Interpreted this way the precautionary principle is misguided. A progress experiment permits us to take the preferable option of proceeding with caution.

Suppose that popular support for genetically modified crops in the United States and Austria remains today as it was in the 1997 survey. These figures would tend to suggest that the United States belongs in the arm of the trial that tests genetic modification as a variety of technological progress. Austria may belong more naturally to the arm of the trial that tests alternative varieties of progress in agricultural technology. For the purposes of a progress experiment we need not inquire too deeply into the reasons individual Americans and Austrians offer to justify their views. It may be that, if pressed, many Austrians would say that genetically modified food is dangerous because its modifiers have introduced DNA into it, and DNA is poisonous. It may be that the optimism of many Americans is grounded in a naïvety about the motives of agricultural biotechnology firms. A progress experiment requires variation. The difference in attitudes between Austrians and Americans shows how that variation could emerge through democratic processes.

This progress experiment does not test the safety of specific varieties of genetically modified organisms. Rather, it tests the viability of an avenue of progress in agricultural technology that involves genetic modification. The agricultural scientists of nations that pursue the path of genetic modification will seek to prevent or minimize its dangers. In those regions in which these crops are not planted there will be pressure to make use of alternative technologies in an attempt to match or surpass the accepted therapeutic and environmental advantages of genetically modified foods. We should not foreshorten this experiment. The experiment should run for a sufficient period of time to ensure that we can detect potential progress traps.

If years pass and benefits accrue to Americans without any indication of a civilization-ending pestilence or reductions of biodiversity or other significant harms, then that tends to support not just the goodness of particular genetically modified crops, but the idea of genetically modified agriculture in general. It would be unrealistic to expect no problems to arise in respect of genetically modified organisms. But if there is a relatively prompt and satisfactory response to each problem then advocates of genetically modified food will confidently cite David Deutsch—they will insist that problems are inevitable but always soluble. The progress experiment will look to a variety of indices of success or failure. It will consider economic benefits and penalties, effects on the natural environment, and so on. Europeans will be in a position to switch to genetically modified products. Their choice will be considerably more informed than the one that the advocates of genetic modification would like to force on European farmers and consumers now. They will have the benefit of an experiment that tested an entire avenue of technological progress, one that included a wide variety of ways of genetically modifying a wide variety of crop species.

It's possible that the future will render a different verdict. Deprived of the option of genetic modification, European agricultural scientists discover other technologies capable of enhancing their crops. These alternative technologies effectively leapfrog genetic modification. European consumers enjoy ample quantities of non- genetically modified food as their newspapers inform them of a succession of American agricultural scandals in which horizontal gene transfer creates super-weeds that destroy a large percentage of cultivable land.

Trials of individual genetically modified crops within nations face the problem of contamination in which a genetically modified organism being tested in one field spreads to neighbouring fields in which non-genetically modified crops are being grown. This problem of accidental spread is likely to be less of a danger in a trial that separates entire nations. Many nations have strict agricultural quarantines designed to protect their agriculture from diseases. The same measures could be used to exclude genetically modified crops from nations in the other arm of the genetic modification progress experiment.

As with the nuclear power experiment we need to let the experiment run for a sufficiently long period. We should not be unduly influenced by some benefit or harm that may not be truly representative of genetic modification as a variety of progress in agricultural technology.

Is it just naïve to think the WEIRD nations that chose correctly would help out those nations who make a mistaken choice? Those nations that reject the path of genetic modification may seem to have spurned a significant commercial advantage. We must remember that there is a significant shared interest that should mitigate the intensity of the competition between nations whose crops are genetically modified and nations whose crops are free of genetically modified organisms. Nations on both sides of this issue should understand that should their bet be mistaken, nations whose agriculture has taken the alternative path will provide a model for them to adopt. They should be prepared to offer their agricultural wisdom should their bet be vindicated. As with the nuclear power example, the experimental approach should help those who have placed the winning bet in respect of the genetic modification of crops to recognize the contribution of the societies that placed the losing bet. The conclusion that genetic modification is a good path for agriculture to follow depends crucially on data gained from societies that reject genetic modification.

The future of technological progress

The idea that we should conduct experiments in technological progress may be difficult for radical optimists to accept. This is partly because they have an inflated view of the value of technological progress. The penalties of falling behind in the development of well-being technologies are less severe than they suppose. Their enthusiasm blinds them to progress traps. The view about the value of technological progress described in Chapter 4 is conducive to the experimental approach. Technological progress improves well-being. However, these improvements are not of the degree supposed by radical optimists. Seeing the goods of technological progress for what they are shows how we can pursue them in a way that does not invite civilization-wrecking disaster. We may have no need to hastily pursue an avenue of progress that initially presents as offering great benefits. We should not view ourselves as akin to the researchers on the Manhattan Project committed to a crash research programme whose purpose was to provide a nuclear bomb as soon as possible to avoid the hideous scenario in which Adolf Hitler got one first.

I think that the downsizing of the benefits of technological progress recommended by this book permits us to return to an older view of the relationship between technology and culture. A visit to a museum with good exhibits of technologies from before the age of colonization reveals a wide variety of well-being technologies. These distinct technologies play a significant role in defining cultures. For example, before extensive European contact in the 1800s the Maori developed distinctive technologies for turning flax plants into clothing, greenstone into adzes and chisels, cooking methods involving heated rocks, and so on. These technologies were the products of a distinctive way of understanding the human relationship with nature and how to use that understanding to sustain and improve well-being. Maori well-being technologies are instantly distinguishable from the well-being technologies that a museum visitor will find in a nearby Incan, or Ancient Egyptian hall.

The progress experiments recommended in this chapter could restore some of that diversity. Reducing the benefits from technological progress should slow the rush to acquire the tools of the technologically dominant culture. Different cultures will be guided by their particular values to explore different avenues of technological progress. Their different sensitivities to and tolerances of risk will direct them to engage differently with humanity's many technological progress experiments.

To some readers this approach will seem to grant undue credence to superstition and other non-scientific prejudices that should be exposed and opposed at every opportunity. But this is not so. Consider a culture whose members oppose agricultural biotechnology because they view its techniques as offensive to nature's guardian spirits. What attitude should those convinced by scientific explanations of nature, explanations that grant no place to guardian spirits, take to the beliefs of this culture? I think that they should see this culture as offering insurance against a progress trap. The dissenting culture need not be seen as advancing claims about the natural world that should be placed on a par with testable scientific hypotheses.

We can be confident that the scientists of future ages will not explain the biological world in terms of guardian spirits. We should be less confident that some of today's agricultural biotechnologies do not combine to constitute a progress trap whose proper understanding is beyond the reach of early twenty-first-century science. If respect for spiritual entities leads some societies to reject certain agricultural biotechnologies then we gain some protection against a progress trap. This is so even if the proper explanation of the trap appeals not to guardian spirits but instead to the posits of some future genetic or ecological science.

Note that the above points apply even if we are confident that there is no hidden progress trap in agricultural biotechnology. It can be prudentially rational to insure your home against destruction by fire even if such an event is properly viewed as very improbable. When considering the purchase of an insurance policy you should consider the badness of outcome that you might insure against and the cost of insuring against it. The destruction by fire of an uninsured house can be a disaster for its owners. This makes it rational to acquire a reasonably priced fire insurance policy even if the insured-against event is very improbable. A lesson from history is that avenues of technological progress about which we feel the greatest confidence can harbour traps obvious in hindsight but invisible in prospect. Permitting, and indeed, encouraging some societies to follow avenues of technological progress that differ from those preferred by WEIRD societies can count as cheap insurance. All we require of the societies that pursue the divergent path is that they have some disposition to amend their views over time. If, after a suitable period of time, agricultural biotechnology brings great benefits without any sign of a progress trap then they should be more disposed to acknowledge it as safe. It is not so important how societies that formerly rejected the practices of agricultural biotechnology represent this acknowledgement. What matters more is that the members of their communities now have access to well-being technologies established as beneficial and less likely to generate a progress trap.

Concluding comments

This chapter has offered a proposal about how best to advance technologically. Early twenty-first-century industrialization could be a globe-spanning progress trap. I point to one response to this danger enabled by a downsizing of the value of technological progress. We should not see ourselves as competitors in some race in which all of the prizes go to the societies that are the most technologically advanced. We can make use of cultural differences in attitudes towards new technologies to conduct experiments in technological progress. I suggest how this might work in respect of nuclear power and genetically modified crops.

6

WHY TECHNOLOGICAL
PROGRESS WON'T
END POVERTY

The radical optimists are confident that they can solve the problem
of poverty. Their general strategy is easily summarized. Poverty is
a consequence of scarcity. Poor people don't have enough food,
shelter, clothing, or other things crucial to lives with good levels
of well-being. The most direct response to poverty is to replace
scarcities with corresponding abundances. The radical optimists
plan to achieve this by the application of powerful technologies to
food, shelter, medicine, and the other necessities of life.

This chapter refutes the technological instrumentalism about
poverty that insists that we best help the poor by promoting
technological progress. Our focus on subjective well-being shows
that the radical optimists have a shallow understanding of poverty.
The low levels of well-being experienced by the poor are not best
understood as consequences of material scarcities. Rather, such
scarcities are, I will argue, typically symptoms of a deeper problem,
a problem that cannot be solved by using technology to generate
material abundances. Indeed, the pursuit of material abundance

by means of technological progress is likely to aggravate it. The errors of the radical optimists become especially apparent when we consider the pivotal role social context plays in judgments people make about how their lives are going. Humans are social animals. Our judgments about how our lives are going are informed by assessments of where we stand in relation to other people. Those who have very little in a world of plenty internalize a negative message about their worth. Herein lies the danger of technologically mediated abundance. The prioritization of technological progress is likely to exacerbate the problem of poverty by enlarging the distance between the haves and have-nots. In its proper place, technological progress can play an important, though strictly adjunct role in our response to poverty. It must not be the predominant response to poverty.

Poverty and well-being

It is apparent that poverty significantly depresses human well-being, and furthermore that this is one of the main reasons for doing something about it. Suppose that we had good reason to think that there was no systematic difference in the well-being of the poor and the wealthy. The defining sentiment of the people of sub-Saharan Africa in the early twenty-first century would be 'poor but happy'. If this was the case we might justifiably view differences in wealth as akin to differences in the ability to play contract bridge, differences in hair colour, and other examples of harmless human variation. But this is manifestly not the case. Poverty significantly reduces well-being.

It might seem that Chapter 4's conclusions about hedonic normalization undercut the idea that we should do something about poverty. If the ancient Romans become hedonically normalized to their age's primitive well-being technologies, then why shouldn't poor people become hedonically normalized to poverty together with its restricted access to the well-being technologies of the early twenty-first century? The evidence suggests that the poor are not hedonically normalized to poverty. They are not stupid. Poor people are aware that they inhabit a world that has more than enough to meet their needs. The fact that people become some-what accustomed to being poor does not eliminate the feeling that more should be done for them. We have some capacity to hedo-nically adjust to the circumstances of poverty—poverty probably does feel worse for someone who finds herself suddenly poor than it does for someone who has grown up with it. But this does not eliminate the effects of poverty on the subjective well-being of those who are born into it. They recognize that there is something wrong about their circumstances. They observe others who are ostensibly no more deserving than they but who have ample supplies of the necessities that they lack.

This chapter's discussion assumes that we have a moral obliga-tion to reduce poverty. We should not stand idly by.[76] The issue I address concerns how we meet that moral obligation. The tech-nology bias suggests one way to meet this obligation. I recommend an alternative.

Ordinary and emergency circumstances of poverty

Before I make my case against radically optimistic solutions to poverty it is important to make a distinction. We should distin-guish *emergency* from the *ordinary* circumstances of poverty.

Technological solutions are more appropriate for the former than they are for the latter.

Consider an emergency case. Someone in the process of starving to death has an urgent need for food. Something must be done now. In this emergency scenario, the end of food is more important than the means by which it is provided. It matters little whether you feed the starving person from your own store of food, thereby making a sacrifice, or whether you effortlessly deploy some technology that causes the needed calories to abruptly materialize. However, emergencies are not the ordinary circumstances of poverty. It is wrong to think of the poor as constantly on the point of death from starvation, disease, or exposure to the elements.

Rich people experience emergencies too. When an avalanche strikes a ski chalet frequented by the rich and famous, its occupants need immediate help. If the avalanche has wiped out their supplies of food they need emergency rations. Their injuries demand prompt attention from emergency physicians. Such emergencies are not the ordinary circumstances of wealth. A wealthy avalanche survivor would presumably object to the suggestion that emergency food rations that she receives immediately after the avalanche should replace her familiar foods in perpetuity. Emergencies are supposed to be, by their very natures, temporary. Some of their effects may be permanent, but we expect a time when the emergency has passed and things will get back to normal. What's appropriate in emergency circumstances is not appropriate when the emergency has passed. So too, the poor are entitled to complain when treatment appropriate to an emergency threatens to characterize their ordinary circumstances.

This chapter's discussion addresses the ordinary circumstances of poverty. In its 2013 report the World Bank reported that there were 1.2 billion people who satisfied the conditions of extreme poverty, living on less than US$1.25 a day. An additional 2.7 billion were classified as vulnerable, surviving on between US $1.25 and US$4.00 a day.[77] Both categories contain people who lack many of our defences against natural disasters. Some of them inhabit war-torn lands. But we acquire too narrow an understanding of poverty if we seek to understand the poor as constantly on the verge of starving to death, dying from disease, or drowning in a flood. Poverty profoundly affects the day-to-day existences of people who suffer it. It matters a great deal how we meet the needs of people oppressed by the ordinary circumstances of poverty. Predominantly technological solutions can suffice to address emergencies suffered by the poor, but they are insufficient when directed at the ordinary circumstances of poverty. An over-reliance on technology threatens to make the poor worse off.

Radically optimistic solutions to poverty

The radical optimist vision of ending poverty by technological means is most readily intelligible as directed at the shortages in the material conditions of human lives. These shortages are frequently cited in complaints about poverty. Technological advances will replace scarcities of food, safe drinking water, health care, shelter, and education with corresponding abundances. Byron Reese sums up the general approach in a way that shows that his calling probably isn't song writing. His riff on the immortal Janis Joplin lyrics 'Freedom's just another word for "nothing left to lose"' is 'Scarcity is just another word for "we don't know how to get it".'[78] Recent presentations of radical optimism offer many accounts of how we can get much more of almost any valuable thing that is currently scarce. They range from general and programmatic accounts of how technological progress can create plenty, to accounts of how specific shortages can be remedied.

Towards the more programmatic end of the spectrum of radically optimistic proposals to end poverty is K. Eric Drexler's proposal of 'atomically precise manufacturing' (APM).[79] Drexler presents APM as a technology that enables us to become 'really good at making things'.[80] APM builds things from the atomic level up. Drexler imagines a car built from feedstocks of relatively simple molecules and atoms. These are formed into generic microscale building blocks and thence into a wide range of larger building blocks. At the end of these iterations we have an atomically precise car. If we are seeking to imagine the world that APM may create Drexler urges that we set the 'base level' for our imaginative exercise 'somewhat above a world-wide abundant supply of the best of every kind of product manufactured today'.[81] He says 'The revolution that follows can bring a radical abundance beyond the dreams of any king, a post-industrial material abundance that reaches the ends of the earth and lightens its burden.'[82]

Currently, poor people are the victims of a vicious cycle. Poverty restricts access to education, the very thing that should enable them to escape from poverty. Their lack of education excludes the poor from the lucrative knowledge economy. According to Drexler, this will be no obstacle to poor people equipped with cheap, atomically precise means of manufacture. He envisages the poor simply manufacturing themselves out of poverty. They will download the plans for the latest atomically precise manufactories and simply press the print button to receive near limitless supplies of any objects they desire. 'Today one can produce an image of the Mona Lisa without being able to draw a good circle; tomorrow one will be able to produce a display screen without knowing how to manufacture a wire.'[83] This strategy should apply to almost any item on the list of material scarcities suffered by the poor. APM will produce sufficient quantities of stuff required to satisfy all of the educational, medical, and other needs of the poor.

Diamandis and Kotler offer more specific descriptions of ways in which material shortages can be replaced by abundances. They say that technologies already in the pipeline will bring drinkable water to the around a billion people who currently lack it. Technological progress will enable the replacement of environmentally destructive methods of food production. Diamandis and Kotler cite Winston Churchill's enthusiasm, in a 1932 speech, for technologies that will permit the manufacture of meat. Said Churchill, 'Fifty years hence, we shall escape the absurdity of growing a whole chicken in order to eat the breast or wing, by growing these parts separately under a suitable medium.'[84]

Today the production of meat is expensive and destructive to natural ecosystems. The quantity of calories consumed by cattle makes beef production a cause of hunger rather than a solution to it. Meat is currently beyond the reach of many people and, from an environmental perspective, it is good that it remains so. Technologies that will enable meat to be grown from stem cells should bring environmentally friendly, humanely produced beef to anyone who wants it. The steaks of the future won't come from environmentally destructive, expensive-to-maintain herds of cattle. They will be grown from beef stem cells. In today's knowledge economy the lack of educational resources available to the world's bottom billion is a significant obstacle. Progress in communication technologies will enable a digitally delivered universal education.[85] The poor suffer significant shortages in healthcare. Diamandis and Kotler expect advances in artificial intelligence and robotics to bring robo-nurses that cheaply and efficiently meet the needs of patients who cannot afford human nurses.[86] I think you get the picture.

It is important, when we confront a problem, to understand what kind of problem it is. The radical optimists treat the ordinary circumstances of poverty as a problem eligible for a technological solution. According to them, poverty is relevantly similar to other problems that have technological solutions. The problem of poverty is relevantly similar to the problem of powered flight solved by the invention of the aeroplane. It is evident that giving food to people who are hungry is part of what must be done. But the prescriptions of the radical optimists omit a crucial ingredient. They fail to adequately acknowledge the importance of people who are currently poor. The recognition of importance is a key feature of the ways in which human beings relate to each other. Healthy relations between human beings involve a mutual respect, a mutual recognition of importance. The radically optimistic approach to poverty is likely to retard this recognition.

The recognition of importance is a feature of all human relationships. We acknowledge deficiencies in such relationships when this recognition is lacking. Consider the complaint expressed by the children of wealthy, career-obsessed people that their parents did not spend sufficient time with them. The complaint is not that the material needs of the children are not met—they typically are. The complaint is that an apparent refusal to make career sacrifices sends the message that the child matters less than the career. A neglected child of wealthy parents does not have her needs met by the purchase of an expensive toy. The toy sends no message about the importance of the child to the parent. A preparedness to make some kind of career sacrifice sends this message.

The relationship between a parent and a child differs greatly from the relationship between rich and poor people. But both are relationships between human beings that require a recognition of importance. The sacrifices you should be prepared to make for your child will tend to be greater than those you are prepared to make for strangers who happen to be poor. But the general point about good and bad ways for humans to relate to each other remains. Some ways of responding to poverty indicate a recognition of the importance of poor people. Other ways do not. A willingness to make sacrifices to improve the circumstances of the poor is evidence of a recognition of this importance.

When a society commits tax revenues to pay the salaries of nurses to meet the needs of elderly poor people it recognizes those needs as important—sufficiently important to warrant spending substantial sums of money. Furnishing 1,000-dollar robo-nurses does not, by itself, acknowledge the importance of those patients and their needs because the recognition of worth implied by the sacrifice is much less. Something beyond the robo-nurses is required.

Were there poor people in the Pleistocene?

There is a tension between the view that poverty is a consequence of material shortages and our understanding of the psychological and emotional mechanisms of subjective well-being as biologically adaptive. To see why, we need to place subjective well-being in its proper evolutionary context. We have supposed that the mechanisms of subjective well-being evolved in the Pleistocene. To be adaptive they must have permitted people living in the Pleistocene to have lives with adequate levels of subjective well-being. At many times during the Pleistocene, humans must have experienced privations of their most essential biological requirements. They must frequently have faced the threat of starvation, risked death from exposure to extreme weather, been sickened by chronic illnesses, been disabled by injury, and so on. There is indirect evidence for the frequency and severity of these hardships in estimates of human life expectancy throughout the Pleistocene. Investigation of the skeletons of Pleistocene humans seems to indicate a very short life expectancy. For example, a study of Late Pleistocene human remains seemed to reveal 'a dearth of older individuals' where 'older' is defined as forty plus.[87] A lack of individuals older than forty would indicate a Late Pleistocene life expectancy—the number of years it would be rational to expect for a Late Pleistocene human newborn—lower than forty.

Some opponents of progress seek to present the Pleistocene in idyllic terms. This is to overlook the very many hardships that technological progress has removed from human lives. Many people in the Pleistocene were one failed hunt away from starvation. Injuries suffered during those hunts, easily treatable by modern medicine, would have led to permanent disability. In the Pleistocene, significant disability is likely to have been swiftly followed by death. Merely shifting a Pleistocene newborn forward in time to the rich world of the early twenty-first century would more than double its life expectancy.

These observations notwithstanding, the material shortages experienced by people in the Pleistocene surely did not prevent many of them from having lives with good levels of subjective well-being. The psychological and emotional mechanisms of subjective well-being would not have evolved in the conditions of frequent material scarcity that characterized the Pleistocene if they had not enabled meaningful engagement with that scarcity. Early humans who fell into extreme listlessness after a couple of failed hunts are unlikely to have survived and passed on their genetic material. Had Pleistocene humans been capable of understanding and responding to life-satisfaction surveys, many are likely to have reported that their lives were going well. Had they reflected on their experiences they are likely to have reported that many of them were positive. To deny the claim that the conditions of the Pleistocene permitted high subjective well-being is to render inexplicable the evolution of the psychological and emotional mechanisms of subjective well-being under the conditions of the Pleistocene. The psychological and emotional mechanisms of subjective well-being would have been maladaptive had misery been the near universal experience of humans in the environment of evolutionary adaptedness.

We might object that Pleistocene humans did not know any better. But the fact that they lacked awareness of conditions in the early twenty-first century explains how they could have high subjective well-being in these circumstances. They were hedonically normalized to the Pleistocene. As we have seen, the fact that it must have been possible to experience adequate levels of subjective well-being in the conditions of the Pleistocene does not mean that tens of thousands of years of technological progress and improvements to the ways in which humans organize themselves into societies are all for naught. The psychological and emotional mechanisms of subjective well-being could have promoted fitness in the environment of evolutionary adaptedness while still leaving a greater percentage of humans in misery than currently experience it.

People living today in the West African Republic of Sierra Leone, one of today's most deprived nations, achieve life expectancies superior to the humans of the Pleistocene. The current life expectancy of people living in Sierra Leone is forty-seven years. That is low by rich world standards. Today, life expectancies in the rich world are in excess of eighty. The difference in life expectancies between the people of present-day Sierra Leone and the people of the Late Pleistocene suggests that the material needs of the former are better met than were those of the latter. It's reasonable to suppose that the people of the Late Pleistocene survived on fewer calories than are available to the inhabitants of present-day Sierra Leone. The life-threatening injuries and diseases of Sierra Leoneans are likely to be more effectively treated than were the injuries and diseases of the Late Pleistocene. The dwellings of present-day Sierra Leone are likely to stand up better against the vagaries of nature better than did the shelters of the Late Pleistocene.

Suppose that the material conditions—access to food, shelter, clothing, medicines, etc.—of people in extreme poverty today are as good or better than conditions typical of the Pleistocene. This comparison suggests two alternative hypotheses about the subjective well-being of poor people.

Hypothesis One: material conditions are all that matter for subjective well-being. People experiencing extreme poverty today should, on average, have levels of subjective well-being at least as good as those of people in the Pleistocene, the environment for which the psychological and emotional mechanisms of subjective well-being evolved.

Hypothesis Two: material conditions matter for subjective well-being, but they are not all that matter. A difference between people in extreme poverty today and people of the Pleistocene explains why people who experience extreme poverty today have levels of subjective well-being, on average, inferior to those typical of inhabitants of the Pleistocene environment of evolutionary adaptedness.

This chapter endorses Hypothesis Two. It explores the contributor to subjective well-being omitted from a focus on the material conditions of people's lives. This missing contributor explains why poverty today consistently produces levels of subjective well-being that were lower than those typical of the Pleistocene. Poor people today lack something of significance that would have been a standard feature of the environment of evolutionary adaptedness.

How poverty affects life satisfaction

We saw in Chapter 1 that there are two measures of subjective well-being. The affective measure is informed by a life's frequency or ratio of positive and negative experiences. The life-satisfaction measure tracks judgments we make about how well our lives are going. These measures reveal two ways in which poverty can depress well-being.

We should expect negative experiences to feature more prominently in the lives of the poor. They have a reduced access to medicines that treat disease. Hunger feels bad. Persistent hunger can cause disease, resulting in additional negative experiences. Some positive experiences are likely to be less available to the poor. The parents of poor children are unlikely to be able to provide the latest PlayStation consoles.

We should expect poverty to have a different kind of effect on life satisfaction. A quick inspection of the contents of your conscious mind should suffice to tell you whether you're currently experiencing pain or pleasure. Judgments about how well our lives are going seem, in contrast, to depend essentially on social context. They are comparative. There doesn't seem to be a variety of feeling you can readily access that tells you that, all things considered, your life is going well, badly, or indifferently. Equally, a list of your activities purged of information about their social context seems insufficient. We are social animals acutely interested in the doings of others. Witness the universal human interest in gossip. It's difficult to imagine a measure of how well a human life is going that does not depend on some form of comparison. Our lives go well or badly in relation to the lives of members of our social circles, the lives of our parents, or the lives of people we follow on Twitter.

The tools furnished by technological progress connect differently to these measures of well-being. We can eliminate some of the unpleasant experiences of hunger or disease by discovering technologies that provide more food or more medicine. These indicators of ill-being are straightforwardly susceptible to remedy by replacing scarcities with the appropriate abundances. However, manifestations of ill-being registered by the life-satisfaction measure are less accessible to technological progress. If such judgments are essentially comparative then abundances could make things worse if the poor receive them, but in quantities far inferior to those received by the rich.

The poor of the wealthy societies of the early twenty-first century possess TVs, cars, hot running water, and other things that rich people of past ages would have lacked. In spite of this, many of them experience hardship and misery. This trend could continue into the material superabundance of the future as imagined by radical optimists. Reece writes excitedly about a future in which progress in manufacturing permits a Mercedes Benz to be built for $50.[88] Be assured that, if and when that day comes, the Mercedes Benz automobile will no longer be something that the wealthy use to distinguish themselves from the poor. It will lose much of the value that it currently possesses. Mercedes Benz automobiles will have travelled the path taken by aluminium cutlery. The most honoured guests of the emperor Napoleon III in the middle of the nineteenth century dined with cutlery manufactured from aluminium. This was before aluminium could be cheaply smelted. Aluminium is highly reactive, making deposits of aluminium metal rare. The advent of the electrolytic process of extracting aluminium from its oxide did not turn all of those who were able to acquire aluminium cutlery into wealthy people. It destroyed aluminium cutlery as a marker of wealth. The technological progress that brings $50 Mercedes Benzes should have a similar effect.

I suspect that the major harms of poverty today correspond to the life-satisfaction measure. Remember that your level of life satisfaction is a matter of the judgments you make about how your life is going. People may experience significant physical pain and yet still say that their lives are going well. Today's wealthy commit themselves to many projects that cause unpleasant sensations. They spend sleepless nights working out how to make the next career move. They join expeditions to climb Everest predictably suffering the sensations of exhaustion and oxygen depletion that come as one ascends the world's highest peak. These pains notwithstanding, the wealthy are more likely to experience high levels of life satisfaction. Comparisons with others reveal that they are doing well. The poor of affluent societies may have televisions that bring to them the pleasures of *American Idol* and *The*

Simpsons. But their comparisons enforce in them the recognition that they are not doing well.

Support for the idea that the relationship between haves and have-nots matters more than the material wealth of the have-nots comes from recent work by Richard Wilkinson and Kate Pickett.[89] The members of wealthier but less equal societies seem, in general, to experience lower levels of well-being than do members of less wealthy, but more equal societies. Their chosen measure of inequality between nations compares the income of a nation's top 20 per cent with the income of its bottom 20 per cent. They say:

> The problems in rich countries are not caused by the society not being rich enough (or even by being too rich) but by the scale of material differences between people within each society being too big. What matters is where we stand in relation to others in our own society.[90]

The incidence of health and social problems is quite closely correlated with a society's degree of inequality and not particularly well correlated with its average income. The more unequal your society the more likely you are to suffer mental illness, be obese, die prematurely, be a victim or a perpetrator of crime, and so on.

The comparisons relevant to life satisfaction must be to people with whom one stands in some meaningful social relationship. There are many such meaningful social relationships. Co-worker, neighbour, fellow citizen, Facebook friend, and follower on Twitter are examples. Wilkinson and Pickett focus on relationships between citizens of a given country. We are significantly socially related to our fellow citizens in a way that we are not related to foreigners. But social relations do nevertheless extend beyond national borders. Poor people in unequal societies are acutely aware of the disadvantages they suffer in relation to the rich of their society. The global poor are also aware that they inhabit a world in which some people have great wealth. The forces of globalization and a sense brought by the climate crisis that we are all connected to a single, fragile environment tends to meaningfully interconnect every human being alive today. We recognize that our decisions are relevant to them and theirs are relevant to us.

Misunderstanding the happiness of the Sun King

Some comparisons fail to inform life satisfaction because they do not correspond to meaningful social relationships. In his demonstration of the huge benefits of technological and economic progress Matt Ridley compares life in an early twenty-first-century liberal democracy with the life experienced by a historical figure whose wealth and luxurious circumstances awed his contemporaries—Louis XIV of France, known as the Sun King. Louis' primary residence was the fabulous palace of Versailles. Ridley describes the Sun King dining alone each evening choosing from forty prepared dishes served on gold and silver plates. Four hundred and ninety-eight people were involved in the preparation of each evening repast.

Ridley proposes that the economic and technological progress that separates the Sun King's time from ours makes him unworthy of envy. Ordinary folk today have it better than the Sun King. 'The cornucopia that greets you as you enter the supermarket dwarfs anything that Louis XIV ever experienced.'[91] Ridley continues:

> You may have no chefs, but you can decide on a whim to choose between scores of nearby bistros, or Italian, Chinese, Japanese or Indian restaurants, in each of which a team of skilled chefs is waiting to serve your

family at less than an hour's notice...You have no wick-trimming footman, but your light switch gives you the instant and brilliant produce of hardworking people at a grid of distant nuclear power stations.[92]

It gets better for the ordinary folk of the technologically advanced early twenty-first century. Runners were available to send the Sun King's missives to any part of his kingdom, but we have the greater reach and near instantaneous service of mobile phones and email. The King's private apothecary lacked most of the medicines available in a local pharmacy. And so on.

This comparison is part of a case for radical optimism. The technological and economic progress that has occurred between the time of Louis XIV and our time has given ordinary people well-being superior to that of France's best life in 1700 CE. The same processes of enrichment should continue.

A focus on the determinants of life satisfaction reveals the irrelevance of Ridley's comparison. Ordinary inhabitants of the rich world of the early twenty-first century may have access to material resources that were beyond the Sun King. But there is a fact about him that does not apply to us and could not come to apply as a result of technological progress. Louis XIV was entitled to anticipate the sentiment expressed by his descendant, Louis XVI, in Mel Brooks's film *History of the World: Part One*—'It's good to be the king'. To be the king is to stand in a special kind of relationship to many people who play the role of subjects. People stopping by their local supermarket on the way home from work do not stand in this relation to the owners and employees of the supermarket. To test this claim out, next time you visit a supermarket try commanding the beating of the manager of the delicatessen section whose slow service disappoints you.

In terms of some measures of material wealth, early twenty-first-century office workers may seem to be doing better than the Sun King. But this is an entirely idle comparison. Donald Trump, an iconic member of the early twenty-first-century moneyed elite, could be doing badly when compared with people of a future age whose fabulous technologies permit them to take holidays on the moons of Jupiter and instantly cure incipient cancer. The early twenty-first-century office worker's realization that she can choose from a wider range of cuisines than those available to the Sun King has no effect on her life satisfaction. And nor should Trump's well-being be adversely affected when he is informed that, in material terms, he is worse off than the poor people of 2300. Both cases involve comparisons between people who are not relevantly related. The awareness that you lack access to well-being technologies to which others alive today have ample access reduces your life satisfaction. Understanding that you, as someone alive in the early twenty-first century, lack access to the well-being technologies of 2300 does not affect the judgments you make about how well your life is going.

Evidence from status competitions for the relevance of social context

Why should the gaps between the haves and have-nots matter so much? There is a plausible evolutionary explanation for the relevance to well-being of social context. The wasteful jockeying for positions on status hierarchies towards the top end of the wealth pyramid has been well documented. There has been less interest in status competitions at the bottom end of the wealth pyramid. I propose that coming last in such competitions brings a special form of vulnerability that is especially injurious to well-being.

Humans competing in status competitions seem to invest much effort in pursuit of things that should matter very little. You purchase a Porsche 911 only to discover that your neighbour, colleague, or romantic competitor owns two of them. You make considerable sacrifices to ensure that your second European sports car is a Ferrari F12 Berlinetta thus leapfrogging your competitor. Viewed from the perspective of satisfying material wants this seems wasteful. It seems to involve viewing Porsches and Ferraris as profoundly different from A-to-B vehicles, a description that surely comes quite close to fully describing their designed functions. Social psychologists who discuss this phenomenon tend to focus on past reproductive benefits for those who emerge triumphant. They say that males who finish first acquire significantly greater reproductive opportunities than do males who finish second even if there is no difference in the satisfaction of basic biological needs.

Here is a conjecture about the evolutionary penalties of finding yourself towards the bottom end of status hierarchies. There is an evolutionary consequence of coming last that explains why those at the bottom of the social heap tend to experience a significant reduction of well-being. Those towards the bottom end of status hierarchies experience an especially dangerous form of vulnerability. Humans during the Pleistocene formed groups whose size could not have been too much greater than a hundred individuals. It would have been very important to remain connected with the other members of your group. There is unlikely to have been much tolerance of those who were perceived not to be contributing towards the collective good. Pleistocene communities did not have much opportunity to accumulate the wealth that would have permitted them to support non-contributors. Evolutionary psychologists find evidence for this great importance in our fascination for gossip.[93] Those who were oblivious of what was going on in a hunter-gatherer band risked finding themselves facing exile.

Today's world differs from the environment of evolutionary adaptedness. We do not systematically exile or kill the socially marginal. But it is nevertheless reasonable to suppose that a psychology designed for the conditions of the Pleistocene should be especially sensitive to this danger. If you experience persistent hunger in an age of plenty then you receive a message that you don't really belong. As the gap between the haves and the have-nots gets greater the more emphatically that message is sent. Having slightly less than others in a fairly egalitarian hunter-gatherer band does not send the same message of dispensability as that received by those at the bottom of less egalitarian social arrangements.

How do Sierra Leoneans of today and humans of the Late Pleistocene differ? The Sierra Leoneans have an awareness of how they stand in relation to others lacked by people of the Pleistocene with similar material needs. It is apparent to them that they live in a world of plenty. We see evidence of this awareness in clothing emblazoned with Western corporate logos. Those who wear t-shirts with images from Disney's *Toy Story* movie franchise know about the material largesse that permits frequent holidays to Disneyworld. A single TV in a poor village presents to its viewers a world of material largesse that contrasts starkly with the conditions of their own existences.

Many of the 3.9 billion people classified by the World Bank as either living in extreme poverty or vulnerable are aware that they live in a world in which others enjoy great wealth. Wealthy people have enough to not only satisfy their basic biological needs, but also to satisfy some quite extravagant wants. Hedonic normalization does not blind the poor to this. A failure to satisfy basic biological needs under these circumstances sends a quite different message from that received by people in the Late Pleistocene. Malnutrition in the Late Pleistocene does not correlate with low importance—it was a near inescapable aspect of human life during that time. It does today. It indicates a widespread judgment that people living in extreme poverty have low moral worth and therefore low entitlements to basic necessities.

Economic and technological trickledown

This does not mean that it is wrong for aid agencies to alleviate material shortages. They should do so. But there are more and less effective ways to address scarcities if your goal is to enhance the well-being of the poor. This chapter's approach to poverty suggests that material scarcities are symptoms of a more fundamental shortage. The needs of the poor are not treated as if they are sufficiently important. This lack registers in the life satisfaction of poor people. It is apparent in the attitudes of the wealthy to both their own nations' poor and to the global poor. If material shortages are symptoms rather than the true causes of the misery of poverty then there will be more and less effective ways to address them. When we offer a predominantly technological solution to the problem of poverty we focus on these symptoms in the wrong kind of way. Our response fails for reasons analogous to the failure of treatments for diseases that target symptoms and not causes.

Predominantly technological solutions are unlikely to increase the well-being of the poor. It is useful to compare the reasons for this failure with the reasons for the failure of another approach to poverty. According to advocates of economic trickledown, the best way to enhance the economic positions of poor people is to permit the rich to enhance their own positions. When wealthy people do better they tend to act in ways that improve the circumstances of the poor. They buy products and services creating jobs in the manufacturing and service industries. Economic trickledown can be implemented globally too. We look to poor parts of the world to supply cheap labour. Work in factories producing athletic shoes and computer motherboards gives us things that we want at a price that we find cheap while funnelling money into the poor world. A focus on the number of dollars in people's pockets would seem to suggest that everyone wins. But this is mistaken. To the extent that well-being is sensitive to social context, we should expect these measures to make things worse rather than better. The most significant benefits of economic progress go to those who are already wealthy. It tends to make the gap between the haves and the have-nots wider. To the extent that the problem is generated by the magnitude of the gap, economic progress can make the problem of poverty worse.

Predominantly technological solutions to poverty can be compared with the solution of economic trickledown. We can call this technological trickledown. The poor are not among the immediate beneficiaries of technological advances. There is a delay before they can afford the new smartphones and cancer therapies. But technological progress tends to make them more widely available. The sooner the wealthy receive them, the sooner they become available to the poor. It follows that the rich and poor have a shared interest in expedited technological progress. This reasoning is mistaken if you believe that the problem of poverty is, in large part, a consequence of the magnitude of the gap between the haves and have-nots. Technological progress in societies like ours tends to give the biggest rewards to the wealthy. It threatens an enlargement of the gap between rich and poor. As a consequence, it exacerbates the problem of poverty. We narrow the gap by judging the poor sufficiently important to warrant a sacrifice by the rich. We fail to rectify it if we grant the necessities of life in a way that indicates no willingness to make a sacrifice.

Consider how the issue of technological trickledown plays out in the important domain of education. We currently face a problem of meeting educational needs. One reason that the children of the poor are unable to compete against the children of the rich is that they are less well educated. There is a global shortage of teachers. In figures published in 2013, the United Nations Educational, Scientific and Cultural Organization (UNESCO) predicted that, globally, an additional 6.8 million teachers were required by 2015 to make good on its commitment to give every child a teacher.[94] This goal is unlikely to be met.

The radical optimists propose a technological solution. Advances in information technology have enabled a One Laptop Per Child initiative whose aim is to provide 'every child on the planet with a rugged, low-cost, low-power, connected laptop'.[95] The aim is a price tag of US$100 per laptop. These and other technological innovations enhance the capacity of students in the poor world to seek their own educational ends.

No one should doubt the potential value of the One Laptop Per Child initiative. It could connect the children of the poor to the wealth of knowledge on the Internet. Future advances that turn One Laptop Per Child into One Robo-teacher Per Child could bring enormous educational benefits. But it is a mistake to think that these technological advances suffice to solve the problem. Educational shortages are not a technological problem with a technological solution. To see this, consider a goal of education that is likely to be set back by treating educational shortages as a technological problem.

One of education's most significant functions is to prepare young people to participate in society. Ideally the children of the poor and rich compete equally for society's rewards. There is a long history of failing to achieve this goal. Societies that do not educate girls, or give them educations that suit them only for domestic duties, contain few women in leadership roles. Herein lies the danger of treating education as a technological problem. Technological progress has the potential to widen and not narrow the gap between the children of the haves and the children of the have-nots. Suppose that technological progress continues to boost the quality of education. One thing that we can predict is that the children of the rich will benefit from technological novelties earlier than the children of the poor. It's true that technologies tend to become cheaper and therefore more widely available once they are introduced. But there is a delay. The logic of exponential improvement suggests that the gap between the rich and poor will get wider rather than narrowing. Were 2030's children of the poor to be competing for jobs against the 2010's children of the rich then we might expect the former to have an advantage. However they will be competing against 2030's children of the rich. The logic of exponential improvement suggests that the gap between 2030's rich and poor children will be still greater than the gap between 2010's rich and poor children.

The preceding paragraph does nothing to gainsay the contribution that exponentially improving technologies can make to education. But we should not treat the shortcomings of our current educational system as technological problems with technological solutions. A traditional response to the unmet educational needs of the poor involves some form of sacrifice in which educational resources paid for by the rich are allocated to the poor. We train teachers and send them to parts of the world that lack teachers. This sacrifice brings an additional benefit of narrowing the gap between the children of the poor and the children of the rich.

Concluding comments

This chapter has investigated the basis of the misery of poverty. I argue that it is wrong to view this misery as a direct consequence of material shortages straightforwardly remediable by technological advances. Material shortages matter in today's world because of what they indicate, a reduced moral significance. It sends the message that poor people don't really matter. The low status experienced by people who have little in a world in which others manifestly have enough to satisfy not only their needs, but also their most extravagant wants, significantly depresses their subjective well-being.

7

CHOOSING A TEMPO OF
TECHNOLOGICAL PROGRESS

The radical optimists make two mistakes about the effects of technological progress on well-being. First, they wrongly infer the long-term pattern of improvement of well-being from the improvement of well-being that individuals experience in their own lives. Hedonic normalization prevents the intergenerational transmission of many of the improvements of subjective well-being that result from advancing technologies. Second, radical optimists subscribe to an instrumentalism that significantly overstates the contribution of technological progress to other important goals that we collectively pursue. We should not seek to invent our way out of poverty or injustice. As we saw in Chapter 6, there are no technological solutions to poverty. Approaches that treat poverty as a predominantly technological problem are likely to exacerbate it.

Rejecting radical optimism does not compel acceptance of a Luddism that sets the value of technological progress at zero. Technological progress boosts well-being. We may adapt completely to certain aspects of a new smartphone (e.g. the clean lines of its design and the variety of its ringtones), but we don't adapt completely to other aspects (e.g. the access it permits to social networking sites and the opportunities to learn by regularly consulting Wikipedia). The fact that hedonic normalization prevents the intergenerational transmission of a large part of these benefits does not purge them from existence. Moreover, incomplete normalization permits some of the hedonic benefits produced by new technologies to carry over to successive generations. Instrumentalism overstates the importance of technological progress. But it is nonetheless clear that new technologies make significant contributions to many of the things that we seek as a society.

The Luddites are wrong. But so are the radical optimists. This leaves an expanse of possible ways to think about the value of technological progress. The purpose of this chapter is not to recommend a precise pace of technological progress, but rather to offer two ideas that should guide decisions about its pace.

The specifics of well-being technologies are not the focus of this chapter's discussion. Technological progress is a temporal phenomenon that occurs when technologies are relevantly better than their precursors. You describe progress in mobile phone technologies not by listing all of the cool things your iPhone can do, but instead by describing the difference between its attributes and the attributes of the devices whose development gave rise to it.

My concern here is in how we go about choosing among a range of possible *tempos* of technological progress. To choose a tempo of progress that is slower than one we might have chosen is not to reject any particular technologies. Rather, it is to choose to receive some technologies later than we might otherwise have received them. These decisions should take account of the predictable effects on subjective well-being of different tempos of progress (see Fig. 9).

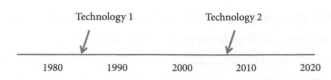

Slower tempo of technological progress

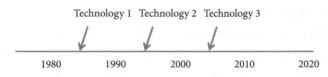

Faster tempo of technological progress

FIG. 9 Faster and slower tempos of technological progress.

According to the radical optimists, we best enhance collective well-being by ensuring that progress in well-being technologies gets as close as possible to the tempo characteristic of information technology. The major selling point of exponential progress is the very fast tempo it achieves when its curve enters the phase of rapid growth. The radical optimists expect this tempo to deliver technologies whose propensities to improve well-being increase in magnitude.

I argue that increases in the tempo of technological progress deliver diminishing marginal contributions towards subjective well-being. By this I mean that faster tempos of progress reduce the propensity for each individual technological advance to boost well-being. This pattern of diminishing marginal value has practical implications for societies seeking to find the right trade-offs between support for technological progress and support for worthwhile collective ends properly viewed as distinct from technological progress. Societies should begin to trade off increases in support for technological progress against support for competing goals. Goals that compete with technological progress for our support include enhancing social justice, equalizing educational opportunities, reducing unemployment, and many more. The faster the tempo of progress, the greater the likelihood that we will correctly judge that other significant collective goals should take precedence over promoting technological progress.

Is there a tempo of progress that it is especially important for a society to achieve? This chapter concludes with a claim that we should try very hard to achieve a *subjectively positive* rate of technological progress. Technological progress is subjectively positive when it is apparent to most users and observers of technology. This tempo conveys a general sense that our technologies are becoming more powerful. Subjectively positive progress in well-being technologies sends a message that new technologies will address some of a society's outstanding problems.

Comparing different tempos of progress

In Chapter 2 I argued that there is no inconsistency in saying that: (1) technological progress conforms to a law, and (2) that different choices by the users and inventors of technology can influence its pace and direction. The law of exponential technological progress is conditional. It assumes certain choices in respect of technology. This assumption is appropriate; humans have had a persistent interest in inventing new technologies. But the law does not prevent us from deciding differently. A community of Luddites does not falsify the law of exponential progress by taking no interest in inventing new technologies. It merely fails to satisfy a condition for exponential progress. I am not saying that this would be a good decision. In choosing not to satisfy any of the conditions assumed by a law of technological progress the Luddite community would deny itself the hedonic benefits described in Chapters 3 and 4. Fortunately we can avoid this Luddite extremism by resolving to partially satisfy conditions for exponential technological improvement. That way we can get the benefits of technological progress while minimizing some of its risks.

Here are two ways in which one can think about the value of increasing the tempo of technological progress. One suggestion comes from the radical optimists. Additional increments of technological progress have increasing marginal value (see Fig. 10). As the line depicting improvements of well-being technologies becomes steeper the value of those improvements increases.

I argue for a different view. Increases in the tempo of technological progress make diminishing marginal contributions to well-being (see Fig. 11). Suppose that a faster tempo of technological progress fits a greater number of new well-being technologies into

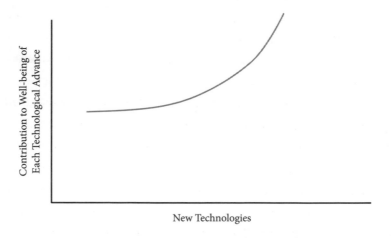

FIG. 10 New technologies make increasing marginal contributions to well-being.

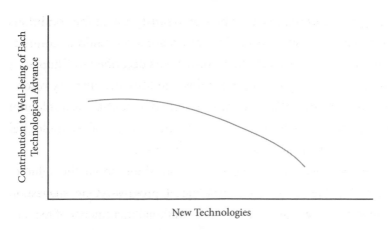

FIG. 11 New technologies make decreasing marginal contributions to well-being.

a given time period than does a slower tempo. The faster tempo tends to reduce the hedonic value to us of each individual technological advance. I propose a mechanism that explains this pattern. This does not imply that technological progress ever comes to have a negative value. We can allow that it is always good to invent new, more powerful well-being technologies. It means that over a given time period, additional well-being technologies tend to make less significant contributions to collective subjective well-being.

Why should such a conclusion be significant? Suppose that technological progress produces returns of increasing marginal value. There are significant implications for trade-offs between technological progress and other goals that do not exhibit the pattern of increasing marginal value. We will, in general, do more to enhance well-being by committing our efforts and resources to goals that make increasing marginal contributions to well-being than when we commit them to goals that do not conform to this pattern. There is, for example, no reason to think that our attempts to enhance social justice produce increasing marginal contributions to well-being. Social injustices do not become systematically easier to rectify the greater the number of them that we have addressed. An interest in maximizing collective well-being would seem consistently to direct our efforts towards the promotion of technological progress.

If technological progress has diminishing marginal value we should witness a weakening in the strength of the motive to pursue it. There are certain kinds of cost associated with progress that do not tend to decline with increases in its tempo.

There are direct costs of technological progress.[96] Some new technologies are socially disruptive. The introduction of computers into the workplace was bad news for those adapted to pen, paper, or typewriter. Skills that took years to acquire and hone lost much of their value. Advocates of technological progress celebrate the process of creative destruction associated with the economist Joseph Schumpeter.[97] The 'destruction' part of this process may be painful, but it sweeps away inferior skills and technologies, permitting new, superior skills and technologies to take their places. We end up better off than we were, or would have been but for the new technologies. This is all well and good, but we should avoid passing a universal judgment on the forms of creative destruction that result from the introduction of new technologies. We should allow that in many cases, perhaps as a general rule, the benefits of technological creation tend to outweigh the losses caused by technological destruction. With a bit of support and training many people adapted to the introduction of computers. Some of those who couldn't were content to go into retirement. We should, however, expect there to be exceptions to this general rule—technological advances whose costs outweigh their benefits. If progress's hedonic returns have diminishing marginal value then we might predict that this would be the case when there is a great deal of change crammed into a relatively short period of time.

There are opportunity costs associated with progress. If accelerations of the tempo of progress predictably bring diminishing marginal returns then there will predictably be a point when resources and effort that could be dedicated to further accelerating the progress of technology are better committed to the pursuit of competing goals. The resources consumed by technological progress take many forms. Most obviously, there is the money invested by government, private corporations, and individuals in developing new technologies. We should add to this, money spent on applied science courses in schools and universities. If part of the value of pure science—the enterprise of predicting and explaining the world—comes from its contribution to applied science, which involves making practical use of that information, then some fraction of resources committed to pure science should be viewed as invested in technological progress. When a society allocates these resources elsewhere it chooses a slower tempo of technological progress than might otherwise have been achieved.

The instrumentalism of the radical optimists encourages us to view these alternative ends as best sought by the introduction of new technologies. We should expect problems out of the reach of today's technologies to succumb to tomorrow's technologies. The faster the tempo of progress the sooner we should expect technological solutions. But, as we have seen from our investigation of poverty and social injustice, instrumentalism is false.

Technological progress makes diminishing marginal contributions to well-being

Economists understand that money has diminishing marginal value.[98] A gift of $100 tends to increase the well-being of a poor person to a much greater extent than it increases the well-being of a rich person. The gift permits the poor person to do many things that would be impossible without it—it might pay for much-needed clothes for children; it may help to repay a loan accumulating crippling interest; it may permit a minor but essential repair for a car, and so on. The gift does little for the rich person. Her wealth permits her to buy many of the things she desires. Doubtless she desires some things that are beyond her current means, but it is very unlikely that $100 would put those things—apartments in Monaco, Ferraris, first-class air tickets to attend the finals of the next Grand Slam tennis tournament—within reach. The richer you are to start with, the less the gift should tend to boost your well-being. Viewed from the perspective of poverty, the $100 gift is more valuable than when viewed from the perspective of wealth.

This fact about the diminishing marginal value of money has implications for plans to boost well-being. Suppose that the only means you had to benefit the poor people of your community were gifts of $100 banknotes. You can hand over the banknotes and seek to measure the effect of your gifts on the recipients' well-being. The diminishing marginal value of money should lead you to expect successive gifts to, in general, have a declining effect on well-being. There will be thresholds that lead some poor people to benefit more from a later gift than from an earlier one. Suppose your most pressing need is for car repairs that will cost $300.

The third gift will boost your well-being to a greater degree than the first two. But in a sufficiently large sample of poor people such thresholds will be distributed across a range of monetary values.

There is a temporal equivalent of diminishing marginal value. It assumes the same basic facts about the psychological and emotional mechanisms of well-being. According to this temporal version, if we receive goods too close together in time they tend to increase subjective well-being to a lesser extent. We maximize the effects on subjective well-being of a collection of goods by temporally spacing them out. This pattern applies to technological progress. A slower pace of progress permits us to receive greater hedonic benefits from each individual new well-being technology.

I will illustrate the general idea of a temporal equivalent of diminishing marginal contributions to human well-being by means of a simple example. Consider how the temporal spacing of birthday presents influences the value that you place on them. Suppose you were to receive, for your next birthday, five years' worth of birthday presents. In this scenario, each of the attendees at your birthday celebration arrives bearing five presents rather than one. They do not cheap out. Each of the gifts considered in isolation would be perfectly acceptable as their sole gift. On the day of your birthday, you allow hourly intervals to space out the opening and enjoyment of each round of presents. In the first hour, you open the first of each of your guests' presents. In the second hour, you move on to the second of their gifts. Viewed purely in terms of the number of gifts you receive this could be the best birthday of your entire life. But there's one sense in which it would be disappointing. This abrupt largesse reduces the value to you of each individual gift. The value to you of each gift would be enhanced if the presents were spread over five years of birthdays rather than being crammed into one. This more gradual schedule would also increase their combined contribution to your well-being.

Consider the effects on your well-being of an extrapolation of this trend. Imagine a version of your birthday in which each guest brings not one or five, but instead one hundred gifts. You receive these at five-minute intervals throughout the day. Limitations on your capacity to enjoy presents mean that this birthday probably doesn't feel all that much better than the five-gift version. Overindulgent parents learn that there quickly reaches a point at which adding extra gifts of toys does nothing to boost a child's mood.

The spacing of birthday presents manifests the temporal equivalent of diminishing marginal value. We maximize the combined value of the presents by spacing them, freeing up time to dedicate to the enjoyment of each. The question we need to answer concerns whether the invention of new well-being technologies conforms to this pattern. Suppose that we restrict our focus to a set number of new well-being technologies. Can their combined effect on well-being be increased if we choose to increase the temporal intervals between them? I suspect that the answer to this question is often yes. The combined value of a set number of technological novelties can increase with greater temporal spacing. Just as we maximize the hedonic value of five birthday presents by spacing them, so too we maximize the value of five advances in well-being technologies by sufficiently spreading them out in time. Receiving technological advances too quickly tends to significantly reduce the hedonic value of each individual technological advance.

Mobile phones and cancer therapies

Consider this point about diminishing marginal value in the context of the rapidly advancing technology of the mobile phone. Information technologies have made significant contributions to the development of mobile phones. To use the language introduced in Chapter 2, mobile phones were infected early in their existence by the rapid exponential improvement of information technology. This has produced a quick succession of improvements. The first mobile phones, introduced in the early 1980s, were bulky, had short battery lives, and were not particularly reliable. Since that time, there have been many improvements. Mobile phones have become more portable. They have become more reliable. They have acquired a variety of new features ranging from text messaging, access to the Internet, access to GPS satellites, cameras that take high-quality photos which can then be sent from one device to another, a variety of software applications that permit a mobile phone to be customized to the particular needs and interests of users, touch screens, and so on. Many of today's mobile phone users retain vivid memories of the hedonic rewards of each advance.

Suppose the pace of progress in the well-being technology of the mobile phone had been more rapid than it in fact was. The gaps between the advances of text messaging, predictive messaging, access to the Internet, and so on would have been shorter than they in fact were. It seems likely that we would have witnessed a phenomenon similar to what occurs when birthday presents are insufficiently spaced. The combined hedonic value of advances in mobile phone technology is greater if they are adequately temporally spaced. The hedonic value of a particular advance (e.g. the

177

advance that permitted a mobile phone to send text messages by means of its number keys) is reduced if it is soon trumped by another significant advance (e.g. touch screens that permit the use of the qwerty keyboard to send messages).

The example of advances in mobile phone technology may seem to trivialize technological progress. It's fun to be able to use your phone to take a 'selfie' and near instantaneously send it to a friend. But these advances seem to lack the great import of advances in medicine, for example. It's one thing to choose a tempo of technological progress that predictably delivers advances in mobile phone technologies more slowly than otherwise. It seems quite another to consciously choose a tempo that delivers improvements in cancer medicine more slowly than otherwise. Cancer kills. Choosing to receive significant advances in cancer medicine later than you might have received them predictably accepts the deaths of people who could have been saved.

It is important to remember that the claim about diminishing marginal value is a claim about the *relative* value of advances specific to a particular lineage of technologies. It states that, in a given technological lineage, the contribution of each successive technological advance to well-being, however great it is, will tend to decline, the more of them that we receive within a given time period. This pattern applies equally to lineages of well-being technologies that make great contributions to well-being and to lineages that make comparatively minor contributions.

I argued that an acceleration of the tempo of progress in mobile phone technologies tends to reduce the hedonic value of each advance. The fact that advances in cancer medicine boost well-being to a much greater extent does not prevent the same pattern obtaining here too. Consider two advances in our treatment of a particular cancer—say lymphoma—of approximately equal magnitude. These advances increase the well-being of people affected by lymphoma, whether as patients, as friends and loved ones of people who suffer lymphoma, or merely as people who live in fear of contracting lymphoma. If advances in that lineage make diminishing marginal contributions to well-being then the first of these advances should boost well-being to a greater extent than the second. This claim is perfectly compatible with the suggestion that the second advance in the treatment of lymphoma boosts well-being to a very great extent.

Were we to possess a magic wand capable of curing cancer, then it should be waved at the first opportunity. We possess no such wand. We must seek the cure by committing resources and effort. There is no way to accelerate progress in medicine that is free of costs. Accelerating progress in medicine predictably involves allocating fewer resources and less effort to other worthy goals. Eisenhower understood that each additional rocket fired imposed a cost in terms of 'those who hunger and are not fed'. There is a point at which we, as a society, should worry more about unfed and uneducated children than we do about finding additional support for cancer research. If we acknowledge that accelerations of progress in cancer research tend to produce diminishing marginal improvements of well-being it becomes apparent how this can happen.

The suggestion that there is an upper limit of support for a certain kind of progress does not imply any view about where we currently stand in relation to that upper limit. The fact that we should trade off advances in cancer therapy against other ends does not show that the level of support for cancer research is currently too high.[99] The sense that we are barely advancing against cancer chips away at the sense of well-being of anyone who occasionally reflects on the potential for one of the trillions of cells that compose her body to abruptly begin to unstoppably divide. I will have more to say about the effects on collective well-being of our awareness of technological progress. We should refrain from taking money essential to the provision of early childhood education to boost the tempo of progress against cancer. But we also spend money on ends that are comparatively unworthy. Doubtless readers will have their own list of unworthy ends that consume public moneys. My personal list of ends that do little to enhance collective well-being includes the use of public moneys to pay large bonuses to underperforming bank executives and resources invested in weapons of war. It is not that these things are entirely undeserving of public support. Rather it's that the level of support for them is currently too high. Much public money currently spent on bankers' bonuses and new weapon systems would be better spent on other things. I leave for another place the elaboration of arguments for my particular list of projects that do not deserve their current degree of support. The claim that progress in cancer medicine deserves no automatic priority over other worthy collective goals is compatible with the suggestion that it deserves a greater degree of support than it currently receives.

We should not seek the fastest tempo of progress in medicine that it is possible for us to achieve. But we should seek the fastest tempo of progress in medicine consistent with giving due weight to other important priorities. The same points apply to the search for other novel well-being technologies. The radical optimists exaggerate the significance of technological progress and so give it an unwarranted pre-eminence among our priorities. Recognizing that technological progress offers diminishing marginal contributions to collective subjective well-being enables support for technological progress to find its proper place among our collective priorities. We acknowledge that there is a point at which effort and resources committed to other priorities predictably do more to enhance well-being.

The importance of subjectively positive technological progress

We've seen that additional increments of technological progress make diminishing marginal contributions to well-being. There predictably comes a point at which we should prefer to support alternative collective goals. Radical optimism tends to obscure this fact. In this section, I argue that there is a minimum tempo of technological progress that it is important for a society to achieve. It is important for a society to achieve a subjectively positive tempo of technological progress. Failing to achieve subjectively positive progress in an important well-being technology can have serious consequences for collective well-being.

It's clear that new well-being technologies boost well-being in a variety of ways. Are there facts about technological progress itself that positively affect well-being? I propose that the answer to this question is yes. We can distinguish hedonic benefits from technological progress from effects on well-being specific to given technologies.

Consider the example of moral progress. Moral progress occurs when there is an improvement in the moral standards of a given society. It comes in a variety of forms. There is moral progress when a society abolishes slavery. There is moral progress when a nation that formerly denied women the vote adopts universal suffrage. It's clear how abolishing slavery boosts collective well-being. Being treated as the property of another person significantly depresses someone's well-being. The benefits of emancipation more than compensate for hedonic losses suffered by dispossessed slave owners. Are there facts about moral progress itself that positively affect well-being independent of the effects of specific moral advances? I suspect so.

The idea that we are making headway on our society's most pressing moral problems has a positive effect on collective well-being. The news that human slavery persists into the early twenty-first century saddens us both because of the harm it causes to those who are enslaved and because it is a rebuff to our idea of moral progress. Memories of long-ago speeches by Frederick Douglass and William Wilberforce encourage us to imagine slavery as an injustice that we have long superseded. It is depressing to learn that it isn't. Commentators on the current obesity crisis observe that this generation may experience shorter life expectancies than the one that preceded it.[100] This is a rebuff to an idea of progress according to which we give our children lives that are predictably better than our own. One of the really depressing aspects of the climate crisis is that the sum total of our productive activities may be making the world a worse place to inhabit rather that a better one.

I suspect that we can identify a hedonic benefit from techno-logical progress that is separable from the beneficial effects of the technologies themselves. It's probable that our species has always undergone technological progress. A desire to understand our circumstances and to seek improvements in them seems to be a universal feature of humans. A community of humans would undergo zero technological progress only in the most unusual of

circumstances. It is important to distinguish the process of technological progress from our awareness of it. Technological progress
is a universal feature of human societies, but we have not always
been aware of it. There has probably never been zero technological
progress, but through long periods of human history technological
progress was nevertheless *subjectively* zero. Progress is subjectively
zero when it is indistinguishable by most people in a given society
from a situation in which there is actually zero progress. If the lives
of members of societies experiencing subjectively zero technological progress seem to be getting better, the improvements will
not seem to be coming from the invention and introduction of
new technologies.

There was technological progress in ancient Egypt. That much
is apparent to historians interested in improvements in pyramid
construction. This notwithstanding, it is likely that many ancient
Egyptians were unaware of this progress. It will seem to the
members of a community undergoing subjectively zero progress
that the technologies they currently possess are little different to
those possessed by their grandparents. A soldier departing for war
will not feel at any disadvantage if he marches off carrying his
granddad's (properly maintained) sword and shield. Exactly the
same herbs that your grandmother used to treat childhood rashes
present as the best option for today's childhood rashes. The
members of this society should feel justified in assuming that
their grandchildren will have technologies that are no more
powerful than their own. Proper care and maintenance will
allow the soldier to pass his weapons technologies on to his
grandson. Your grandmother's herbal remedies will seem as valuable to your grandchildren.

Some support for the idea that people in the past experienced subjectively zero technological progress comes from the historical popularity of cyclical views of time. Cyclical views of time were prominent in the thinking of the some of the Ancient Greeks, the Hopi, the Mayans, the Babylonians, Hindus, and Jains. A favoured metaphor of these accounts is the wheel. Each revolution of the wheel of time is supposed to return us to our collective starting point. The idea of time reliably repeating itself could only make sense for those who lack any awareness of the directional nature of technological change. In today's age of accelerating technological progress it would be hard to believe in the existence of a natural process capable of unobtrusively eradicating all evidence of the microprocessor and the Airbus A380.

When considered as a whole, it is clear that this is an era of subjectively positive progress. The exponentially improving information technology that drives smartphones and laptops conveys the strongest impression of technological progress. One does not have to be actively engaged in the business of inventing new technologies to have a keen awareness of technological progress. Your grandparents' wedding photos reveal times when people made do with manifestly more primitive well-being technologies. People expect that the future made by our grandchildren will contain well-being technologies significantly more powerful than those that we have today. To be a resident of a prosperous early twenty-first-century democracy is to have different expectations of technological progress from those of the residents of Egypt in 2000 BCE.

But a closer inspection reveals that the impression of technological progress is uneven. There is variation in the perceived pace of progress across different domains. The pace of progress seems rapid in information technology. Today it seems less rapid in cancer medicine.[101] During the early 1970s, people had the mistaken sense that a cure was imminent. The architects of US President Richard Nixon's War on Cancer evidently thought that this war, in contrast with wars waged in South East Asia, was eminently winnable. These hopes were dashed. Siddhartha Mukherjee's book *The Emperor of All Maladies* presents a contemporary understanding according to which cancer is an enemy we will never entirely vanquish. Mukherjee says:

> Cancer is a flaw in our growth, but this flaw is deeply entrenched in ourselves. We can rid ourselves of cancer, then, only as much as we can rid ourselves of the processes in our physiology that depend on growth – aging, regeneration, healing, reproduction.[102]

As we have seen, there is certainly progress in cancer medicine. But the question of whether this progress is *subjectively* positive concerns what the public is aware of. Popular disappointment about progress against cancer derives in part from a comparison between cancer medicine and other areas of medicine in which progress seems to have been rapid. There has been significant and obvious progress against the infectious diseases that killed large numbers of our ancestors. Smallpox has been eradicated. Tuberculosis remains, but is much less of a threat to the inhabitants of today's rich world than it was to their eighteenth-century ancestors. More recently, improvements in diets and reduced rates of smoking have led to significant reductions in cardiovascular disease. When compared with these manifest successes, progress against cancer is bound to disappoint. Moreover, progress against these causes of death has increased the incidence of cancer. The mathematics of cell division and genetic mutation mean that the risk of cancer rises steeply with increases in life expectancy. The older you are the more likely you are to receive a cancer diagnosis. As the seas of infectious illness have receded they have further exposed the reefs of cancer. People today live with an awareness that, at any given time, one of the many trillions of cells that comprise their bodies may be about to begin dividing in an uncontrolled way. Cancer is a bad-luck lottery in which everyone has effectively purchased many trillions of tickets. The perception that we are barely progressing against cancer has a negative effect on subjective well-being.

I propose that subjectively positive technological progress is an important contributor to the well-being of the citizens of techno-logically advanced societies in the early twenty-first century. Our awareness of technological progress engenders in us a realistic expectation that some of today's outstanding problems can be solved by improved technologies. Some of those that cannot be decisively solved will at least have their severity significantly reduced.

The popularity of recent expositions of radical optimism results partly from a near universal need to see our belief in human improvement vindicated. I have challenged the degree of improve-ment forecast by radical optimists. But this does not undermine a general belief in the great value of technological progress. Things can, in general, be getting better without improving exponentially. Rejecting radical optimism about the degree to which techno-logical progress can improve well-being leaves in place the idea that the improvement of our technologies should tend to make things better. Subjectively positive technological progress enhances well-being by creating a sense of optimism that some of today's intractable problems both can be addressed and are worth seeking to address. Individuals who belong to societies that achieve this standard of technological progress are both aware of this progress and have some confidence that it will address some of the problems that matter most to them. I propose this as a minimum standard. If we are in danger of not achieving the standard of subjectively positive technological progress then we should make sacrifices in our pursuit of other goals.

Concluding comments

This chapter has offered some suggestions about how to set the tempo of technological progress. I argue that technological progress makes diminishing marginal contributions to well-being. A society that invests too many resources and too much effort in accelerating technological progress should predictably fall short in other important areas. I propose that we should strive especially hard for subjectively positive progress in important categories of well-being technology. It is apparent that we have already achieved this in some technologies. But we fall short in others.

AFTERWORD

Don't turn well-being technologies into Procrustean beds

I have waited until the final pages of this book to reveal an assumption that I have made about technological progress. I have placed great emphasis on research into how humans respond to positive and negative events in their lives. I proposed that this research restricts the benefits we can expect from technological progress. But there's a way of escaping the limitations imposed by hedonic adaptation and hedonic normalization. This path of escape emerges directly from the accelerating technological progress that has been my principal focus.

Among the bounty brought by technological progress are technologies that should permit humans to modify themselves. Rather than just tolerating the limitations imposed by hedonic adaptation and hedonic normalization we could seek to intervene in the psychological and emotional mechanisms of subjective well-being. The aim would be to boost our capacity to enjoy technological novelties. This is entirely in keeping with our established approach to biological limitations. Humans didn't evolve wings. But we didn't just accept that we should remain earth-bound—we invented technologies to carry ourselves into the air. If there's a feature of our evolved psychology that's preventing us from deriving the maximum enjoyment from technological novelties then why shouldn't we change it?

Our current technological environment is very different from the environment of evolutionary adaptedness. It's therefore not surprising that we are imperfectly attuned to its dangers and benefits. We now recognize the tendency to readily consume all available fat and sugar, useful in an era when calories were scarce, as a cause of today's obesity epidemic. We are seeking ways to modify this evolved feature of human psychology to better suit environments with plentiful cheeseburgers and doughnuts. Why shouldn't we do the same for the aspects of our psychologies that limit our capacity to benefit from more powerful well-being technologies? We might change our human natures to better fit the environments that exponential technological progress is creating.

The most forthright advocates of using technology to modify ourselves are the transhumanists.[103] According to the transhumanists, there's nothing sacrosanct about human nature. If human nature prevents us from getting what we want, then we should use technology to change it for the better. The transhumanist approach seems to be premised on the same can-do attitude that has always driven technological progress. We apply technologies to inconvenient aspects of nature. Why not do the same for inconvenient aspects of our own human natures?

A range of technologies currently under development promise modifications that will enable us to derive greater hedonic rewards from technological progress. There are technologies that should permit the modification of human genetic material. Interventions in human genomes could soon permit the alteration of genes involved in our hedonic responses.[104] Other focuses of research involve cybernetic implants that could be grafted to human brains. These might be programmed to enable humans better to adapt to and enjoy a future of very rapid technological progress. This is the end game of technological progress as imagined by Ray Kurzweil.[105] He views exponential technological progress as taking us towards a future in which we progressively fuse our biological brains and bodies with machines. We will trade in clumsy, disease-prone biology for computer chips and cybernetic prostheses. The accelerating pace of technological progress means that the replacements will soon significantly outperform the biological parts that they replace. Kurzweil presents this replacement as ongoing, leading eventually to an outcome in which biology has totally given way to technology. Humans will, to quote the subtitle of his book, 'transcend biology'. This could be good news for our subjective well-being. The electronic components of our future artificial minds will be programmable in ways that better attune them to technological progress. A programmable human mind may enjoy technological novelties in a way that is not eroded by hedonic adaptation. This undiminishing bliss will not prevent an appropriately programmed mind from seeking additional pleasures.

The transhumanist proposal will not appeal to those who believe that we have souls that are likely to be detrimentally affected by attempts to alter ourselves. I want here to suggest a more pragmatic reason for rejecting this course, one that does not suppose the existence of souls or any other supernatural entities.

The designers of well-being technologies make certain assumptions about human beings. Typically, there is a gap between how human beings are supposed by the designers of technologies to be, and how we actually are. There are many cases of technologies that disappoint because their design is insufficiently informed by the needs and capacities of real human beings. Don Norman, a writer on the cognitive science of technological design, responds to this phenomenon with a call for a human-centred design.[106] This design philosophy 'puts human needs, capabilities, and behavior first'.[107] Among Norman's favourite examples of insufficiently human-centred technologies are sliding doors. The handles of many such doors send a message to human users that they should be pulled rather than slid. According to Norman, practitioners of human-centred design should think carefully about how human capabilities and limitations will incline us to use technologies. Sometimes this is easier said than done. In the case of some technologies—those that assume certain mental capacities—we lack the knowledge required to perfectly fit the technology to us.

A well-being technology based on an understanding of human beings that is almost, but not quite right may often succeed in enhancing well-being but sometimes have the reverse effect. What tends to happen is that these mistakes become apparent, scientific understanding of human beings advances, and new understanding permits the invention of superior well-being technologies. Designers of sliding doors find new ways to make their operation apparent to human users.

Each era has an understanding of human beings that abets its ideals of technological progress. In the 1930s many scientists applied behaviourist theories to explain human beings.[108] Behaviourists sought to understand humans and animals by focusing exclusively on observable behaviour and measurable stimuli. They rejected appeals to unobservable subjective states in the explanations of human behaviour.

This way of understanding human beings informed the search for new well-being technologies. One among these was the language laboratory as a means of instruction in a second language. Language laboratories of the 1950s taught by means of repetition. A language learner would sit in a booth in a language laboratory and hear and repeat samples of language. Someone seated in a booth in a language laboratory could be fed a large number of samples of Japanese. A behaviourist theory of language learning suggests that mastery of Japanese should emerge after repetition of a sufficient number of samples of the language.

Language laboratories still exist. But the demise of behaviourism saw a reduction in their perceived importance in the teaching of foreign languages. We now understand that human beings are not the simple input–output devices that some behaviourists supposed. Modern approaches to second language learning are more cognitive. They give due emphasis to the teaching of an understanding of how a language works.

Recognition of that fact led to new ways to teach second languages. But it is possible to imagine another way to respond to this failure, one that was thankfully not available when behaviourism was the reigning paradigm of human learning. A technology-centred design philosophy could use genetic and cybernetic technologies to modify humans to better fit our technologies. In the case of language instruction we would modify ourselves to better fit our language teaching technologies. We could become superior input–output devices. These modifications would permit us to turn the samples of language presented in a language laboratory into mastery of a language. They would enhance our ability to benefit from language laboratories.

It is now possible to see just how bad an idea this would have been. Behaviourists oversimplified our understanding of language. Language laboratories are not complete failures as language teaching technologies. They can play a useful supplementary role in contemporary second language instruction. They are, however, a well-being technology that is understood by teachers to not conform exactly to the psychologies of human language learners. Our understanding of language would have been detrimentally affected had we sought to conform our psychologies to the requirements of language laboratories.

According to an ancient Greek myth, Procrustes was an innkeeper who claimed to possess a bed whose length exactly matched whoever lay down in it. The catch was that the bed was not magical—it did not lengthen or contract to exactly match the height of any guest. Rather, Procrustes would alter the physical dimensions of his guests to fit the bed. He would remove lower limbs of taller guests and stretch shorter guests. This is how he achieved the perfect fit between guest and bed. The hero Theseus brings to an end Procrustes' career in the ancient Greek hospitality industry by fitting Procrustes to his own bed.

The story of the Procrustean bed is sometimes presented as an objection against social engineers who advance policies that ignore the many differences between people. A state that seeks to standardize education ignores the many differences in the ways children learn. But the story of Procrustes' bed can be told in a way that warns against efforts to redesign ourselves to better fit our well-being technologies. Procrustes' bed is a well-being technology whose design was insufficiently informed about variation in the dimensions of human beings. We should not turn language laboratories—or any other well-being technology—into Procrustean beds. The inexact fit between our well-being technologies and the physical and psychological makeup of human beings means that even the best of them will be imperfectly tailored to our needs.

We are, however, likely to sacrifice things of value if we use the well-being technologies as templates to redesign aspects of our natures.

There are more contemporary examples of disappointments that seem to come from oversimplifications of human beings made by designers of well-being technologies. The practical interests of designers of new technologies necessarily simplify human beings. For the designer of a well-being technology, close enough is typically good enough. As we saw in Chapter 3, there is unlikely to be a perfect fit between social networking technologies and human emotional and psychological capacities. An over-reliance on Facebook seems to amplify feelings of inadequacy and anxiety in some people. The correct conclusion to draw out of such research is not that we should renounce social networking technologies. Rather it is that we should think about which uses of them make us happy and which depress us.

The danger of modifying ourselves to better fit our well-being technologies is that we cannot be sure exactly what we will lose. We are complex beings. We lack full understanding of our own complexity. We know now what would have been lost had we taken a Procrustean approach to technologies that assumed a behaviourist conception of ourselves. Time in a language laboratory sufficient to present us with samples of all of the linguistic elements of Japanese would have taught us the language. But treating the language laboratory as a Procrustean bed would have deleted features of ourselves that we now recognize as very valuable. For example, it is likely to have impaired our capacity to creatively reflect on language and to imagine new ways to use words. We can only guess at what we might lose if we were to alter ourselves to more closely conform to the requirements of today's well-being technologies. There are inductive grounds for thinking that we would lose much. This loss is apparent when we look back and consider how things would have been had past generations possessed the tools to tailor humans to more closely fit their technologies.[109]

This book is no attack on technological progress. Rather, it is an attack on certain overly ambitious ways of thinking about technological progress. We should not expect an exact fit between the technologies we invent and our natures—philosophical and scientific understanding of human nature is likely always to be a work in progress. We should continue to seek the benefits of improved technologies while maintaining a safe distance from them.

ENDNOTES

1. Matt Ridley, *The Rational Optimist: How Prosperity Evolves* (New York: Harper Collins, 2010), p. 352.
2. For a short summary of recent work on subjective well-being see Ed Diener, 'The Remarkable Changes in the Science of Subjective Well-Being', *Perspectives in Psychological Science* 8(6) (2013), pp. 663–6. Influential works in the field include Ed Diener and Martin Seligman, 'Beyond Money: Toward an Economy of Well-Being', *Psychological Science in the Public Interest*, 5 (2004), pp. 1–31; Daniel Kahneman and Alan Krueger, 'Developments in the Measurement of Subjective Well-Being', *Journal of Economic Perspectives* 20(1) (2006), pp. 3–24; Richard Layard, *Happiness: Lessons from a New Science*, Second Edition (London: Penguin, 2011).
3. Prominent philosophical approaches to well-being include Fred Feldman, *What Is This Thing Called Happiness?* (Oxford: Oxford University Press, 2010); James Griffin, *Well-being* (Oxford: Clarendon Press, 1986); Daniel Haybron, *The Pursuit of Unhappiness* (Oxford: Clarendon Press, 2008); Richard Kraut, *What is Good and Why?* (Cambridge, MA: Harvard University Press, 2007); Daniel Russell, *Happiness for Humans* (Oxford: Oxford University Press, 2012); Wayne Sumner, *Welfare, Happiness, and Ethics* (Oxford: Clarendon Press, 1996); Mark Walker, *Happy-People-Pills for All* (Malden, MA: Wiley Blackwell, 2013).
4. Matt Ridley, *The Rational Optimist: How Prosperity Evolves* (New York: Harper Collins, 2010), p. 352.
5. David Deutsch, *The Beginning of Infinity: Explanations that Transform the World* (London: Allen Lane, 2011).
6. Eric Schmidt and Jared Cohen, *The New Digital Age: Reshaping the Future of People, Nations, and Business,* (London: John Murray, 2013), p. 257.
7. Schmidt and Cohen, *The New Digital Age*, p. 23.
8. Byron Reese, *Infinite Progress: How the Internet and Technology Will End Ignorance, Disease, Poverty, Hunger, and War* (Austin, TX: Greenleaf Book Group Press, 2013), p. 6.

9. Peter Diamandis and Steven Kotler, *Abundance: The Future Is Better Than You Think* (New York: Free Press, 2012), Kindle edition, Chapter 1, Section 2, para. 11.

10. Diamandis and Kotler, *Abundance*, Chapter 1, Section 3, para. 7.

11. K. Eric Drexler, *Radical Abundance: How a Revolution in Nanotechnology Will Change Civilization* (New York: Public Affairs, 2013).

12. Dwight D. Eisenhower, 'Chance for Peace', speech to the American Society of Newspaper Editors, Washington, DC, 16 April 1953.

13. Evgeny Morozov's discussion of technological solutionism in his *To Save Everything Click Here: The Folly of Technological Solutionism* (New York: Public Affairs, 2013), p. 5.

14. As W. Patrick McCray says: 'The prevailing belief of technologists . . . is that technology is the solution to all problems'; W. Patrick McCray, 'The Technologists' Siren Song', *The Chronicle of Higher Education*, 10 March 2014, available at <http://chronicle.com/article/The-Technologists-Siren-Song/145107> accessed November 2014.

15. For accessible histories of cancer and cancer medicine see Siddhartha Mukherjee, *The Emperor of All Maladies: A Biography of Cancer* (New York: Scribner, 2010) and George Johnson, *The Cancer Chronicles: Unlocking Medicine's Deepest Mystery* (New York: Knopf, 2013).

16. This chapter focuses on a cause of progress specific to technologies and technological change. Radical optimists draw on other ideas about progress that are not specific to technology. Consider Matt Ridley's interesting claim that it's 'the ever-increasing exchange of ideas that causes the ever-increasing rate of innovation in the modern world'; Matt Ridley, *The Rational Optimist: How Prosperity Evolves* (New York: Harper Collins, 2010), Kindle edition, p. 269. He proposes that ideas for new technologies result from the recombination of ideas that he characterizes as 'ideas having sex'. Biological sex accelerates evolutionary progress by enabling novel combinations of genetic material. In an analogous way, novel ideas arise out the many different combinations of existing ideas permitted and encouraged by free exchanges. 'Ideas are having sex with other ideas from all over the planet with ever-increasing promiscuity'; Ridley, *The Rational Optimist,* p. 270. Ridley gives some examples from technology: 'The telephone had sex with the computer and spawned the Internet'; Ridley, *The Rational Optimist,* p. 270. The idea for the car arose out of a kind of conceptual sex between the bicycle and the horse carriage. Ridley's suggestion tells us something interesting about progress in general. It's often the case that new insights come from the combination of ideas formerly restricted to distinct ideological domains. Consider

the new insights about disease that are coming from an attempt to place them in an evolutionary context, or the advances in the understanding of the mind that came from the applications of ideas from computer science. This book can be viewed as a product of the sexual union of the ideas of technological progress and subjective well-being. The law of exponential improvement is at the centre of this chapter's discussion because it purports to tell us something specific with regard to progress in technology.

17. I take this example from <http://raju.varghese.org/articles/powers2.html> accessed November 2014. For a very entertaining illustration see Adrian Paenza, 'How Folding Paper Can Get You to the Moon', available at <http://ed.ted.com/lessons/how-folding-paper-can-get-you-to-the-moon> accessed November 2014.

18. Erik Brynjolfsson and Andrew McAfee, *The Second Machine Age: Work, Progress, and Prosperity in a Time of Brilliant Machines,* (New York: W. W. Norton, 2014), p. 8.

19. For a recent expression of this view see Michio Kaku, *The Future of the Mind: The Scientific Quest to Understand, Enhance, and Empower the Mind* (New York: Doubleday, 2014).

20. Ray Kurzweil, *The Singularity Is Near: When Humans Transcend Biology* (New York: Viking, 2005), Kindle edition, Chapter 2, Section 6, heading of figure 1.

21. Kurzweil, *The Singularity Is Near*, Chapter 2, Section 5.

22. Kurzweil, *The Singularity Is Near*, Chapter 2, Section 5.

23. Kurzweil, *The Singularity Is Near*, Chapter 2, Section 5.

24. Kurzweil, *The Singularity Is Near*, Chapter 2, Section 5.

25. Colin McGinn, 'Homunculism', *New York Review of Books*, 12 March 2013.

26. See James Le Fanu, *The Rise and Fall of Modern Medicine* (London: Little, Brown, and Company, 1999).

27. David Koplow, *Smallpox: The Fight to Eradicate a Global Scourge* (Berkeley, CA: University of California Press, 2004) p. 21.

28. Available at the website of the Founders Fund <http://www.foundersfund.com/the-future> accessed November 2014.

29. Byron Reese, *Infinite Progress: How the Internet and Technology Will End Ignorance, Disease, Poverty, Hunger, and War* (Austin, TX: Greenleaf Book Group Press, 2013), pp. 68–72.

30. Kurzweil, *The Singularity Is Near*, Chapter 2, para. 1.

31. Ray Kurzweil, *How to Create a Mind: The Secret of Human Thought Revealed* (London: Duckworth Overlook, 2013), Kindle edition, Introduction, para. 11.

32. Kurzweil, *The Singularity Is Near*, Chapter 2, Section 2, para. 8.
33. Kurzweil, *The Singularity Is Near*, Chapter 2, Section 2, para. 8.
34. Kurzweil, *The Singularity Is Near*, Chapter 2. Section 2, para. 9.
35. Kurzweil, *The Singularity Is Near*, Chapter 2, Section 2, para. 8.
36. These paragraphs may seem to suggest that exponential progress is restricted to information technologies. K. Eric Drexler offers some interesting detail on a future programme of atomically precise manufacturing (APM) that seems likely to undergo reflexive improvement. APM builds large objects—like cars—from the atomic level up. The atomic precision here refers to basic operations that produce precise patterns of atoms. Drexler describes how the manufacture of a car could begin with an iterated process in which feedstocks of relatively simple molecules and atoms are formed into generic microscale building blocks. These blocks are then combined in miniature manufacturing plants into a wide range of larger building blocks. These are assembled into still larger components, and so on up until the creation of a car. Advances in APM could, like advances in information technology, be reflexive because the factories that produce atomically precise things are themselves objects of atomically precise manufacture. An advance in the design of atomically precise manufactories permits further advances in the design of atomically precise manufactories. These advances are therefore not mere passive platforms for later advances. Once APM gets properly underway we should expect to see rapid exponential progress; K. Eric Drexler, *Radical Abundance: How a Revolution in Nanotechnology Will Change Civilization* (New York: Public Affairs, 2013).
37. I am certainly not the first person to draw an analogy between technological change and a biological process. Numerous thinkers have offered evolutionary explanations for the growth of technology in general. See, for example, Kevin Kelly, *What Technology Wants* (New York: Penguin, 2011). There are also appeals to biological processes to explain specific cases of technological change. In 2014 there was some discussion in the blogosphere of an interesting paper by John Cannarella and Joshua Spechler that used a model of infectious disease to ground a prediction that the social networking site Facebook would soon lose most of its users. Infections pass through a population when susceptible individuals either die or acquire immunity. Cannarella and Spechler predict a mass abandonment of Facebook by 2017 with the depletion of the pool of potential users. See John Cannarella and Joshua A. Spechler, 'Epidemiological Modeling of Online Social Network Dynamics', available at <http://arxiv.org/abs/1401.4208?context=physics> accessed November 2014. This chapter's model of the acceleration of

technological progress is informed by different aspects of infectious disease from those appealed to by Cannarella and Spechler. I seek to explain accelerations in the power of technologies that come into contact with information technologies rather than their passage through a population of users.

38. Kelly, *What Technology Wants*, Chapter 8, para. 5.

39. Richard Wilkinson and Kate Pickett, *The Spirit Level: Why More Equal Societies Almost Always Do Better* (London: Allen Lane, 2009), p. 3.

40. Maria Konnikova, 'How Facebook Makes Us Unhappy', *The New Yorker*, 10 September 2013. Original article Ethan Kross, Philippe Verduyn, Emre Demiralp, Jiyoung Park, David Seungjae Lee, Natalie Lin, Holly Shablack, John Jonides, and Oscar Ybarra, 'Facebook Use Predicts Declines in Subjective Well-Being in Young Adults', *PLOS One* 8 (2013).

41. Henry Sidgwick, *The Methods of Ethics*, Seventh Edition (London: Macmillan, 1913), p. 48.

42. Roy Baumeister, 'Choking Under Pressure: Self-consciousness and Paradoxical Effects of Incentives on Skillful Performance', *Journal of Personality and Social Psychology* 46 (1984), pp. 610–20.

43. Dean Hamer, 'The Heritability of Happiness', *Nature Genetics* 14 (1996), pp. 125–6.

44. P. Brickman, D. Coates, and R. Janoff-Bulman, 'Lottery Winners and Accident Victims: Is Happiness Relative?' *Journal of Personality and Social Psychology* 36 (1978), pp. 917–27.

45. Philip Brickman and Donald Campbell, 'Hedonic Relativism and Planning the Good Society', in M. Appley (Ed.), *Adaptation Level Theory: A Symposium* (New York: Academic Press, 1971), pp. 287–302.

46. Sonja Lyubomirsky, 'Hedonic Adaptation to Positive and Negative Experiences', in S. Folkman (Ed.), *Oxford Handbook of Stress, Health, and Coping* (New York: Oxford University Press, 2011), pp. 200–24 at p. 202.

47. Ed Diener, Richard Lucas, and Christie Scollon, 'Beyond the Hedonic Treadmill: Revising the Adaptation Theory of Well-being', *American Psychologist* 61 (2006), pp. 305–14.

48. Lyubomirsky, 'Hedonic Adaptation to Positive and Negative Experiences'.

49. Lyubomirsky, 'Hedonic Adaptation to Positive and Negative Experiences', p. 201.

50. Lyubomirsky, 'Hedonic Adaptation to Positive and Negative Experiences', p. 208.

51. Edward Gibbon, *The History of the Decline and Fall of the Roman Empire*, abridged and edited by David Womersley (London: Penguin Books, 2000), p. 82.

52. Matt Ridley, *The Rational Optimist: How Prosperity Evolves* (New York: Harper Collins, 2010), p. 12.

53. Ridley, *The Rational Optimist*, p. 13.

54. See for example, Ed Diener's answer to the question 'Are most people unhappy?' in his FAQ on happiness and subjective well-being, 'A Primer for Reporters and Newcomers', available at <http://internal.psychology.illinois.edu/~ediener/faq.html#happiest> accessed November 2014.

55. Key expositions of evolutionary psychology include Jerome Barkow, Leda Cosmides, and John Tooby, *The Adapted Mind: Evolutionary Psychology and the Generation of Culture* (Oxford: Oxford University Press, 1992); David Buss, *Evolutionary Psychology: The New Science of the Mind* (Boston: Pearson, 1998); Steven Pinker, *How the Mind Works* (New York: Norton, 1997).

56. See for example, David Sloan Wilson, *Darwin's Cathedral: Evolution, Religion, and the Nature of Society* (Chicago: University of Chicago Press, 2002) and Scott Atran, *In Gods We Trust: The Evolutionary Landscape of Religion* (Oxford: Oxford University Press, 2002).

57. Ronald Wright, *A Short History of Progress* (Cambridge, MA: Da Capo Press, 2005), p. 107.

58. Wright, *A Short History of Progress*, p. 8.

59. Wright, *A Short History of Progress*, p. 39.

60. Jared Diamond, *Collapse: How Societies Choose to Fail or Succeed* (New York: Viking Press, 2005).

61. Wright, *A Short History of Progress*, p. 107.

62. Wright, *A Short History of Progress*, p. 132.

63. David Deutsch, *The Beginning of Infinity: Explanations that Transform the World* (London: Allen Lane, 2011), p. 221.

64. Steven Pinker, *The Better Angels of Our Nature: Why Violence Has Declined* (New York: Viking Press, 2011).

65. Wright, *A Short History of Progress*, p. 56.

66. Wright, *A Short History of Progress*, p. 132.

67. For an interesting discussion of the value of an experimental approach to social policies see Timothy Wilson, *Redirect: The Surprising New Science of Psychological Change* (New York: Little, Brown, and Company, 2011).

68. Jared Diamond, *The World Until Yesterday: What Can We Learn from Traditional Societies* (London: Allen Lane, 2012).

69. Diamond, *The World Until Yesterday*, p. 9.

70. Diamond, *The World Until Yesterday*, p. 9.

71. Karl Popper, *The Logic of Scientific Discovery* (London: Routledge, 2002).

72. Mark Winston, *Travels in the Genetically Modified Zone* (Cambridge, MA: Harvard University Press, 2005).

73. The following figures come from the International Service for the Acquisition of Agri-Biotech Applications. See <http://www.isaaa.org/resources/publications/briefs/44/executivesummary/default.asp> accessed November 2014.

74. David Freeman, 'Are Engineered Foods Evil?' *Scientific American* (September 2013), p. 72.

75. The Wingspread Consensus Statement on the Precautionary Principle, available at <http://www.sehn.org/wing.html> accessed November 2014.

76. Philosophers express a variety of views about how to justify a moral obligation to alleviate the suffering of the poor. According to some, the wealthy should accept responsibility for some of the harms suffered by the poor. This is the view of Thomas Pogge. See Thomas Pogge, *Politics As Usual* (Cambridge: Polity, 2010) and Thomas Pogge, *World Poverty and Human Rights* (Cambridge: Polity, 2002). According to Pogge, the citizens of wealthy nations created and maintain a global order that oppresses the poor. They owe restitution as a matter of justice. Other philosophers do not think that it is essential to view the have-nots as harmed by the haves. This is the view of Peter Singer. The mere fact that poor people are unable to satisfy some of their most basic needs entitles them to assistance from those who have more than enough to satisfy their own basic needs. This assistance is owed even if the rich are entirely blameless in respect of the predicaments of the poor. See Peter Singer, 'Famine, Affluence and Morality', *Philosophy and Public Affairs* 1 (1972), pp. 229–43 and Peter Singer, *The Life You Can Save* (Melbourne: Text Publishing, 2009).

77. The World Bank Annual Report 2013, available at <http://web.worldbank.org/WBSITE/EXTERNAL/EXTABOUTUS/EXTANNREP/EXTANNREP2013/0,menuPK:9304895~pagePK:64168427~piPK:64168435~theSitePK:9304888,00.html> accessed November 2014.

78. Byron Reese, *Infinite Progress: How the Internet and Technology Will End Ignorance, Disease, Poverty, Hunger, and War* (Austin, TX: Greenleaf, 2013), p. 108.

79. K. Eric Drexler, *Radical Abundance: How a Revolution in Nanotechnology Will Change Civilization* (New York: Perseus Books, 2013).

80. Drexler, *Radical Abundance*, p. 286.

81. Drexler, *Radical Abundance*, p. 247.

82. Drexler, *Radical Abundance*, p. 286.

83. Drexler, *Radical Abundance*, p. 247.

84. Peter Diamandis and Steven Kotler, *Abundance: The Future Is Better Than You Think* (New York: Free Press, 2012), Kindle edition, Chapter 9, Section 5, para. 1.

85. Diamandis and Kotler, *Abundance*, Chapter 14.

86. Diamandis and Kotler, *Abundance*, Chapter 15.

87. Erik Trinkaus, 'Late Pleistocene Adult Mortality Patterns and Modern Human Establishment', *PNAS* (2011); published ahead of print, 10 January 2011.

88. Reese, *Infinite Progress*, p. 102.

89. Richard Wilkinson and Kate Pickett, *The Spirit Level: Why More Equal Societies Almost Always Do Better* (London: Allen Lane, 2009).

90. Wilkinson and Pickett, *The Spirit Level*, p. 25.

91. Matt Ridley, *The Rational Optimist: How Prosperity Evolves* (New York: Harper Collins, 2010), p. 36

92. Ridley, *The Rational Optimist*, pp. 35–7.

93. See Robin Dunbar, *Grooming, Gossip, and the Evolution of Language* (Cambridge, MA: Harvard University Press, 1998).

94. <http://www.unesco.org/new/en/education/themcs/leading-the-inter national-agenda/education-for-all/advocacy/global-action-week/gaw-2013/glo bal-teacher-shortage/> accessed November 2014.

95. Diamandis and Kotler, *Abundance*, Chapter 14, Section 2, para. 4.

96. I'm grateful to an anonymous reader from OUP for emphasizing this point.

97. See for example, Erik Brynjolfsson and Andrew McAfee, *The Second Machine Age: Work, Progress, and Prosperity in a Time of Brilliant Machines* (New York: W. W. Norton, 2014).

98. For a helpful discussion of the diminishing marginal value of money, see Richard Layard, *Happiness: Lessons from a New Science*, Second Edition (London: Penguin, 2011), pp. 30–70.

99. There are some who do argue that our current level of support for cancer research has negative consequences for other varieties of medical research. For poignant reflections on the difficulty in competing for funding for research on traumatic brain injury against funding for cancer research, see Sian Lawson, 'Cancer Isn't the Only Disease that Needs More Research', *The Guardian*, 29 April 2014, available at <http://www.theguardian.com/commentisfree/2014/apr/29/cancer-isnt-only-disease-more-research> accessed November 2014.

100. See for example, Doug Dollemore, 'Obesity Threatens to Cut U.S. Life Expectancy, New Analysis Suggests', *National Institutes of Health News*,

16 March 2005, available at <http://www.nih.gov/news/pr/mar2005/nia-16. htm> accessed November 2014.

101. See for example Siddhartha Mukherjee, *The Emperor of All Maladies: A Biography of Cancer* (New York: Scribner, 2010), Kindle edition, and George Johnson, *The Cancer Chronicles: Unlocking Medicine's Deepest Mystery* (New York: Knopf, 2013).

102. Mukherjee, *The Emperor of All Maladies*, 'Atossa's War', para. 4.

103. For a valuable resource on transhumanism see the transhumanist FAQ, available at <http://humanityplus.org/philosophy/transhumanist-faq/> accessed November 2014.

104. J-E. De Neve, 'Functional Polymorphism (5-HTTLPR) in the Serotonin Transporter Gene Is Associated with Subjective Well-Being: Evidence from a U.S. Nationally Representative Sample', *Journal of Human Genetics* 56 (2011), pp. 456–9; Mark Walker, *Happy-People-Pills for All* (Malden, MA: Wiley Blackwell, 2013).

105. Ray Kurzweil, *The Singularity Is Near: When Humans Transcend Biology* (New York: Viking, 2005).

106. Don Norman, *The Design of Everyday Things*, Revised and Expanded Edition (New York: Basic Books, 2013).

107. Norman, *The Design of Everyday Things*, p. 9.

108. John Watson, *Behaviorism*, Revised Edition (Chicago: University of Chicago Press, 1930).

109. This question of whether and how we might use technology to modify ourselves is a topic for another book. For my version of that book, see Nicholas Agar, *Truly Human Enhancement: A Philosophical Defense of Limits* (Cambridge, MA: MIT Press, 2014).

INDEX

AHA!

The Moments of Insight that Shape Our World

William B. Irvine

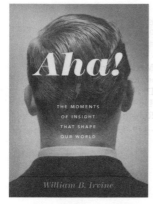

978-0-19-933887-0 | Hardback | £16.99

'Lucid, engaging and thought-provoking'

Irish Mail

Great ideas often develop gradually after studying a problem at length—but not always. Sometimes, an insight hits like a bolt from the blue. In *Aha!* philosopher William B. Irvine explores these epiphanies, from the minor insights that strike us all daily, to the major realizations that alter the course of history. We like to think that our greatest thoughts are the product of our conscious mind. Irvine demonstrates, though, that it is our unconscious mind that is the source of our most significant insights.

Irvine explores not only the neuroscience of aha moments but also their personal and social ramifications. How does a person respond to having a breakthrough insight that goes against a dominant paradigm? And how does the world respond when they share that insight? *Aha!* is a must-read for cognitive scientists, intellectual historians, philosophers, and anyone who has ever been blown away by the ideas that enlighten us when we least expect it.

Sign up to our quarterly e-newsletter **http://academic-preferences.oup.com/**

ELEGANCE IN SCIENCE

The Beauty of Simplicity

Ian Glynn

978-0-19-966881-6 | Paperback | £10.99

'An erudite book...Well illustrated and full of historical anecdote and background, this is an elegant volume indeed.' **Nature**

'There is a wealth of historical information packed in here.' **Times Literary Supplement**

The idea of elegance in science is not necessarily a familiar one, but it is an important one. The use of the term is perhaps most clear-cut in mathematics—the elegant proof—and this is where Ian Glynn begins his exploration. Scientists often share a sense of admiration and excitement on hearing of an elegant solution to a problem, an elegant theory, or an elegant experiment. With a highly readable selection of inspiring episodes highlighting the role of beauty and simplicity in the sciences, this book also relates to important philosophical issues of inference, and Glynn ends by warning us not to rely on beauty and simplicity alone: even the most elegant explanation can be wrong.

FREE

Why Science Hasn't Disproved Free Will

Alfred R. Mele

978-0-19-937162-4 | Hardback | £12.99

'Alfred Mele's beautifully written and easily accessible book is a perfect tonic to the many recent claims by scientists that there is no such thing as free will.'
Michael McKenna, University of Arizona

Does free will exist? There are neuroscientists who claim that our decisions are made unconsciously and are therefore outside of our control, and social psychologists who argue that a myriad of imperceptible factors influence even our minor decisions to the extent that there is no room for free will. According to philosopher Alfred R. Mele, if we look more closely at the major experiments that free will deniers cite, we can see large gaps where the light of possibility shines through. Mele lays out his opponents' experiments simply and clearly, and proceeds to debunk their supposed findings, one by one, explaining how the experiments don't provide the solid evidence for which they have been touted.

HAPPINESS

The science behind your smile

Daniel Nettle

978-0-19-280559-1 | Paperback |
£8.99

'A lucid and sensible survey of the latest research.'
Independent

'Well written, accurate and engaging, with a lightness of touch that makes it a delight to read.'
Nature

What exactly is happiness? Can we measure it? Why are some people happy and others not? And is there a drug that could eliminate all unhappiness? Daniel Nettle uses the results of the latest psychological studies to ask what makes people happy and unhappy, what happiness really is, and to examine our urge to achieve it. Along the way we look at brain systems, at mind-altering drugs, and how happiness is now marketed to us as a commodity. Nettle concludes that while it may be unrealistic to expect lasting happiness, our evolved tendency to seek happiness drives us to achieve much that is worthwhile in itself.

Sign up to our quarterly e-newsletter **http://academic-preferences.oup.com/**

THE FOURTH REVOLUTION

How the Infosphere is Reshaping Human Reality

Luciano Floridi

978-0-19-960672-6 | Hardback | £16.99

'Searing study...non-alarmist and very, very smart.'   ***Nature***

'Fascinating stuff.'   ***New Scientist***

Who are we, and how do we relate to each other? Luciano Floridi, one of the leading figures in contemporary philosophy, argues that the explosive developments in Information and Communication Technologies (ICTs) are changing the answer to these fundamental human questions. "Onlife" defines more and more of our daily activity—the way we shop, work, learn, care for our health, entertain ourselves, conduct our relationships; the way we interact with the worlds of law, finance, and politics; even the way we conduct war.

As the boundaries between life online and offline break down, and we become seamlessly connected to each other and surrounded by smart, responsive objects, we are all becoming integrated into an 'infospher'; a metaphysical shift that represents nothing less than a fourth revolution.

Sign up to our quarterly e-newsletter **http://academic-preferences.oup.com/**